"Asma Hasan will rock your stereotypes about Islam in this refreshing book. Here is a young woman who embraces Islam, modernity, America, her family, and her friends—all with enthusiasm and commitment. She sees no contradictions between them and, after you have read this book, neither will you."

—Fareed Zakaria, author of *The Post-American World*

"Hasan makes for a disarming spokeswoman."

—*USA Today*

"A warm, witty, wonderful story about what it means to be both Muslim and American in a post-9/11 world. This is a book that every American should read."

—Reza Aslan, author of *No god but God* and *How to Win a Cosmic War*

"Hasan's writing is a useful beacon for American Muslims who may be struggling to articulate their identities as both Americans and Muslims."

—Religion

"In her last lively book, As[illegible] m feminist cowgirl'—referrin[illegible] o-rado. In this follow-up work that mixes autobiography with feisty insights into Islam and the many misconceptions people have about it, the author demonstrates the spiritual practice of enthusiasm . . ."

—*Spirituality & Health*

"Hasan's version of Islam would have appealed to America's founders with its advocacy of human equality, religious tolerance, property rights and self-improvement. It harmonizes just as well with 21st-century America's spiritual inclinations: it is nonjudgmental, inclusive, open-minded, diverse, experiential, emotional and even feminist. This is do-it-yourself American religion at its most appealing."

—*Publishers Weekly*

"*Red, White, and Muslim* is a must read for anyone who wants to be inspired by the beauty and wisdom of Islam. With insight, integrity, passion, and eloquence Asma Gull Hasan shares her personal journey as a proud American Muslim and in doing so breaks our stereotypes, melts our fears, nurtures our hopes, and enlightens our minds and hearts. Hasan opens a window into Islam in its American forms and offers us a glimpse of the opportunity and power of Muslim Americans to shape the future of Islam and our world."

—Rabbi Irwin Kula, author of *Yearnings: Embracing the Sacred Messiness of Life*

"With passion and humor and wisdom, Asma Hasan compellingly explains why the seeming contradictions of her life as a traditional Muslim and a modern American are no contradiction at all. Urgent and inspirational, her account helps us understand how American Muslims, much like American Catholics and American Jews before them, are reshaping their religious identity without sacrificing their religious authenticity. Asma Hasan not only loves Islam; she embodies the Islam of love."

—Yossi Klein Halevi, author of *At the Entrance to the Garden of Eden: A Jew's Search for God with Christians and Muslims in the Holy Land*

"Each of the world religions that are at home in the United States has generated a distinct synthesis of its own values and those of American democracy. In each case, the synthesis has eventually had world importance. Islam—old as a religious tradition but relatively new to the United States—is early in this creative process, but the lively and authentic ring of Asma Gull Hasan's personal synthesis proves that the process is definitely under way. At a dark hour, this book is a piece of bright and cheering news."

—Jack Miles, author of *God: A Biography*

"Asma Gull Hasan's, *Red, White, and Muslim* is an engaging book that reminds us that beyond the terrorism headlines, Islam remains a spiritual path for the vast majority of moderate mainstream Muslims that live and work in our societies. The life and reflections of this bright, talented, dynamic young Muslim woman are an excellent example of the fact that for many Muslims today Islam is a source of meaning, guidance and joy in their lives."

—John L. Esposito, professor, Georgetown University and author of *What Everyone Needs to Know About Islam*

"A new generation of Muslim Americans—proud of their heritage and their country—stand poised to take their rightful place at the forefront of American religious and political life. Asma Gull Hasan is an extraordinary spokesperson for this widely diverse community. Honest, disarmingly open, and sparkling with the author's radiance and intelligence, *Red, White, and Muslim* opens a window to a world of faith, reason, and questioning that is distinctively American and Muslim at the same time."

—Noah Feldman, Bemis Professor of Law, Harvard Law School and author of *Fall and Rise of the Islamic State*

"A refreshing book showing Islam through the eyes of a bright Muslim woman in America. Asma is sincere in expressing her own vision with eloquence, integrity, and passion for her beliefs."

—Dr. Maher Hathout, senior advisor to the Muslim Public Affairs Council and Interfaith Alliance Board of Directors member

"Asma passionately shares with us her personal story and through her compelling voice, the reader connects to a convincing universal Muslim story of belief."

—Ranya Idliby, coauthor of *The Faith Club*

red, white, and muslim

red, white, and muslim

My Story of Belief

*

Asma Gull Hasan

HarperOne
An Imprint of HarperCollins*Publishers*

Most quoted Qur'anic passages are from the Muhammad Asad translation. The Abdullah Yusuf Ali translation is quoted alternatively.

Tumhen Dil Lagi Bhul Jani Paregi by Nusrat Fateh Ali Khan. Copyright © 1999 by Nusrat Fateh Translation of the Qur'an from *The Holy Qur'an: Text, Translation and Commentary,* by Abdullah Yusuf Ali. Copyright © 2001 by Tahrike Tarsile Qur'an, Inc. Reprinted by permission of the publisher.

Translation of the Qur'an from *The Meaning of the Qur'an* by Muhammad Asad. Copyright © 1980 by Pola Hamid Asad. Despite best efforts, the copyright holder could not be reached.

"Morning Has Broken" by Eleanor Farjeon from *The Children's Bells* published by Oxford University Press. By permission of David Higham Associates.

Junoon, *Allah Hu.* Music: Salman Ahmad. Lyrics: Traditional. Copyright © 1998 Nameless Music LLC/© 2004 Junoon Incorporated. www.junoon.com.

Translation of Hafiz poems, "I Have Learned So Much," "The Sun Never Says," "Tiny Gods," and "That Lamp That Needs No Oil," from *The Gift* by Daniel Ladinsky. Copyright © 1999 by Daniel Ladinsky. Reprinted by permission of the author.

Translation of Rumi poems from *Rumi: Hidden Music.* Reprinted by permission of HarperCollins Publishers Ltd. Copyright © Maryam Mafi and Azima Melita Kolin 2001.

HarperCollins books may be purchased for educational, business, or sales promotional use. For information please write: Special Markets Department, HarperCollins Publishers, 10 East 53rd Street, New York, NY 10022.

HarperCollins Web site: http://www.harpercollins.com.

Originally Published as *Why I Am a Muslim* by Element in 2004

Library of Congress Cataloging-in-Publication Data is available upon request.

ISBN 978-0-06-167375-7

09 10 11 12 13 RRD(H) 10 9 8 7 6 5 4 3 2 1

Dedication

In memory of my grandfather, Salahuddin Khan,
who I wish had lived to see this book.
For the men in my life,
my father and my brother.
Thank you for always
believing in me, sticking up for me
and being just a phone call away!

Even after all this time
The sun never says to the earth,
"You owe Me."
Look what happens
With a love like that,
It lights the Whole Sky.

—Hafiz

As many Muslims do,
I begin this endeavor with a short prayer
that goes like this:

Bismillah, Ir-Rahman, Ir-Rahim
In the name of God, Most Gracious, Most Merciful

Contents

A Note to the Reader xi
Introduction xv

Chapter 1 Born Muslim 1
Chapter 2 A Direct Relationship with God 17
Chapter 3 Sufism: A Rich Mystical Tradition 39
Chapter 4 We Are All Imperfect 67
Chapter 5 The Diversity of Islam 87
Chapter 6 A Woman's Religion 113
Chapter 7 Being Muslim Makes Me a Better American (and Being American Makes Me a Better Muslim) 141

Bibliography 167
Acknowledgments 169
About the Author 171

A Note to the Reader

It may be that God will grant love and friendship
between you and those whom ye now hold as enemies.
For God has power over all things;
And God is Oft-Forgiving, Most Merciful.

Qur'an 60:7

Dear Reader,

When I was growing up in Colorado in the 1980s, I always saw "Merry Christmas" messages on television around the holidays. I never saw anything for any Islamic holidays, and I took this to mean that people generally did not like Muslims. It didn't bother me; I just accepted it. I was in college in the mid-1990s the first time I saw a "Happy Hanukkah" message on CNN, and I found myself overjoyed at the sight of it. Although the announcement was for a Jewish holiday, I felt liberated somehow—that perhaps it was only a matter of time before the same kinds of messages would appear for Islamic holidays and for other religions, that I lived in a community whose members wished the best for each other.

I hear stories all the time that make me believe this kind of community is possible. My friend Lara teaches high school chemistry in a San Francisco public school. Recently, one of her students was gushing over Lara's full-day fast for the Jewish observance of Yom Kippur. Lara accepted the praises but decided to add, "It's not easy, but it's easier than fasting for a whole month." She continued, "Muslims have to fast for thirty days during Ramadan. Now that's hard, and it shows real dedication." At that moment, a young American Muslim girl in Lara's class, who was sitting toward the back, looked up at my friend with a smile that lit up the classroom. Her teacher was talking about her religion—and in a positive way. Such events so rarely happen, especially these days. Lara called me soon after and told me about her student. We celebrated over the phone. Lara being Jewish and me being Muslim—we see things that affect either of our religions as affecting both of us, that we are in this experiment together, and that we want it to succeed. Lara's small gesture probably made that student's week, maybe even her month. I should know, because I once was that student.

I have never been ashamed to be Muslim, not even after 9/11, and not now. I know that many non-Muslims do not understand Islam but want to learn more. I also know that some Muslims carry out violent acts in Islam's name and use Islam to justify many un-Islamic things. Islam is a major influence in my life. I have been Muslim my whole life, and I could not imagine being anything else.

Although the reasons why I am a Muslim are elementary to me, they are probably not elementary to you. You probably don't know what it's like to be a Muslim girl, growing up in America, to wonder if wearing blue jeans is against Islam, to hope that, this year, CNN will say "Eid Mubarik." But you can gain an understanding of these feelings, like Lara has. You can compare and contrast them to your own experiences as a religious person, a member of a minority group, or as a person with your own unique perspective on life. Maybe you can be on the receiving end of a young Muslim's smile.

I know you are an intelligent person and also curious. Otherwise you would not be reading this book. Do you want to have the knowledge to understand Islam? Do you want to be able to say, "Muslims are not the enemy"? If you think of Muslims as the enemy, eventually they will become the enemy. You have the choice not to predestine your thoughts. This book will give you the tools if you want to use them.

The Islam that I practice is not the one depicted by Osama bin Laden, or by Al Jazeera, cable news, or the fear-mongers. I am not a member of a secret society of terrorists nor do I plot the death of non-Muslims. What Islam is really about is so different from the many misconceptions—about women, about other religions, about even the concept of *jihad.* Islam does not preach violent aggression against one's "enemies." I would not be a Muslim if that were true. In fact, the Qur'an and the core values of American society are strikingly similar. I wrote this book so readers could learn about all these aspects of Islam—and distinguish the facts from the myths. I wrote this book so that non-Muslims will know the things about Islam that only Muslims know—the things that keep me and over a billion people in the world Muslim. My hope is that by the end of my story you will understand why, as an American woman, I proudly am—and choose to be—a Muslim.

—A.G.H.

carry an undercurrent of true feeling. Like Lara, my own complex and [illegible] experiences as [illegible] of a [illegible] [illegible] with [illegible] perspective [illegible] the [illegible] and [illegible] Muslims [illegible]

I know you are an intelligent reader, and [illegible] whatever you would [illegible] really [illegible] knowledge [illegible] [illegible] the [illegible] you [illegible]

[illegible] the death of [illegible] Muslims [illegible] [illegible] that [illegible] Muslims [illegible] people in the [illegible]

Introduction

When I wrote *Why I Am A Muslim* (published in 2004), I felt that I was able to step outside the fray that surrounded Islam and provide an easy-to-read, spiritual description of my reasons for being Muslim. Today, several years later, the "War on Terror," and the focus on Islam, has changed and deepened. I find that non-Muslim Americans and other Westerners are even more challenged in understanding Islam. While some understand the basics (and others still haven't bothered), many ignore them and move on to histrionics and platitudes. Those with good intentions are genuinely confused by the ambiguities of how Muslims actually live their daily lives and form their opinions, which sometimes have no Qur'anic or theological basis. Issues like Sunni-Shi'ite discord, the viability of democracy in an Islamic country, and, most important to me, the status of women in Islam, confuses and terrifies everyone alike, even me as a Muslim.

The pressure on Muslim women now is different than it ever has been. All the issues I mentioned as well as the side effects of the War on Terror come down most strongly on Muslim women. The War on Terror has produced a war on women. When the West asserts itself on Muslim men, Muslim men assert themselves on Muslim women, for lack of any other reachable target. Muslim women, in response, turn off or turn away. I loathe criticizing the Muslim community, but I believe that Muslim women all over the world are facing a spiritual crisis as a result of this pressure. Even for me, constantly responding to "Why don't Muslims condemn terrorism more?" leaves less time for me to nurture and advocate the optimistic and spiritual side of Islam, which is the essence of why I am a Muslim. Everyone who is concerned about the state of the world needs to refocus on and educate themselves about what Islam truly stands for and why Muslims hold the beliefs they do.

Most Muslim women around the world are not conservative in their practice, but moderate. Yet the Muslim leadership, from Wahhabi-American imams to the late (and tragically assassinated) Pakistani prime minister Benazir Bhutto, espouses *hijab*—a conservative concept—as required for Muslim women who practice properly. This issue is sticky for the majority of Muslim women, who do not wear *hijab* (including me). When I was giving radio interviews for the previous edition of this book, in cities across America as well as in London, women with Muslim names would often call in. Without identifying themselves as Muslim, they would ask about some controversial area of Islam—one that Western feminists and critics had picked on, attempting to show Islam's inferiority or its supposed tendency to oppress women. After a few of these calls, I realized that these women were not conservative Muslims trying to challenge me. They were actually moderate Muslims themselves who had been pushed away from the faith and were in disbelief that someone like them still staked a claim for Islam.

"How can you be a Muslim without wearing hijab?" one woman with an Arabic-Muslim name asked me during an interview at an NPR affiliate. Her tone was angry, but I could tell that it was not because I didn't wear hijab, but because she didn't, and therefore felt she couldn't consider herself a good Muslim. What this caller was really asking was: how did I have the guts to consider myself Muslim, even though I didn't subscribe to the establishment views, albeit minority ones?

"How do you explain that, in Islam, women inherit half of what their brothers do?" I was asked once on the air by a bleak and angry female voice. From all these women, I could hear their anger at Islamic practices, their frustration. How could the religion these callers were raised in let them down like this, betray them? I understood their frustration. While on the one hand, our community teaches us all the rights Islam celebrates for women and that are stated in the Qur'an, a conservative minority increasingly touts repressive principles and cultural, tribal attitudes. Most Muslims don't agree with these repressive views, but share the progressive values I describe in this book. When we allow this conservative minority to define our religion, those of us with the correct view of Islam are browbeaten into believing we have the wrong interpretation.

The very identity of Muslim women—even Muslims in general—is in question. Many of us Muslim women don't fit the full conservative model of Muslim womanhood accepted at the mosque, yet we can't bring ourselves to wear a bikini on the beach or eat bacon with our pancakes. I hear this frustration in the voices of fellow Muslim women, a frustration that has markedly increased in the last few years. I feel this conflict within myself. It is challenging to belong to a religion whose male adherents often ignore not only the tenets of the religion, which clearly stand for equality, but also the repression they are exerting on their own mothers, wives, daughters, and friends. The community leadership often says that being a Muslim female means wearing hijab, standing in the back of the mosque and at the back of Muslim burials, even for your own relatives. I am afraid that many Muslim women, possibly millions around the world, are saying in response, "*Fine.* We're not Muslim then." These external, conservative, and actually extra-Islamic markers of being a Muslim woman are leaving many Muslim women out. They don't convert to another religion, but they think of themselves as being secularized, or simply raised Muslim but not much more.

In Ethiopia, where I visited in 2006 as a speaker for the U.S. State Department, I realized how far Muslim women are being held back. The status of women in Ethiopia is abysmally low.

People all over the world, especially Muslims, are eager to hear about American Muslims. I have travelled to Europe, Asia, and Africa and have held many videoconferences with the U.S. embassies of Muslim and non-Muslim countries around the world. Many of the participants are fascinated that Islam even exists in the United States, that there are Muslims who were born into American Muslim families and raised Muslim in America, who speak English without an accent and sound American.

I would be lying if I didn't say that I spent most of the Ethiopia trip on the receiving end of vociferous complaints about the war in Iraq and American imperialism. Most of the State Department officials accompanying me didn't even attempt to defend U.S. policies. Besides the officials, one group at these gatherings was also noticeably quiet: the middle-aged Muslim women sat silent, with blank expressions. In contrast, the Muslim men and younger Muslim women were enlivened by their newly found Wahhabi beliefs. (Wahhabism, a literal and conservative interpretation of Islam, is a rather new development in the Muslim world, especially outside of the Arab world. I am not against different interpretations of Islam, but I feel that Wahhabis and other conservative Muslims want to strangle any interpretations of Islam outside of their own.) Lecturing against the U.S. and modern Muslim women came easily to these young Muslims, but any real dialogue did not. It was ironic to me that in the ancient Ethiopian walled city of Harar, where I was giving a talk, a young, soccer-jersey-clad Wahhabi man lectured me on how my dress and appearance were sinful and wrong (in addition to the hatred he had for the United States). The middle-aged women at the event, who said nothing and avoided making eye contact, grew up with a different Islam. Their faces were those of the shamed victims of a crime, just staring ahead, not giving away any emotion. Every woman wore hijab, as a rule. Even I was tricked into wearing it on one occasion, and on others simply gave up and put it on.

After the soccer-jersey Wahhabi and his cronies left, the middle-aged Muslim ladies approached me. They told me that they disagreed with the Wahhabi views but couldn't speak up. They said that since 9/11, Wahhabism has grown. (It is possible that these Wahhabis, seeing the United States attacked so successfully, have been emboldened, although they had nothing to do with the attack.) This growth in Wahhabism requires that all women wear hijab and not attract attention to themselves in public. If a woman defies these rules, she cannot leave her house at all without being harassed. So they said they have just given in, in exchange for some semblance of a normal life, some hope to still be part of society, to go to lectures like mine and work at least. The Harar women told me that, previously, most Ethiopian women didn't wear hijab. It was a new phenomenon with the growth of Wahhabism. I do not understand how the Muslim world expects to grow and flourish when half its population likely feels imprisoned without bars. A society cannot advance if most participants aren't free to contribute to that advancement.

The only private event I had was with a group of Muslim women parliamentarians. I expected a room full of the Hillary Clintons and Nancy Pelosis of Ethiopia. But it was the opposite. In fact, the women were so modest and mild-mannered that it was hard to believe that they held public office at all. They did their best to blend into the background. I found out during the meeting that most of these women were teachers who had held local posts. When the prime minister won his reelection in 2005, in a hotly disputed contest, he dismissed all elected officials who were not in his party, from the national bodies all the way down to local assemblies, sometimes even imprisoning them. These dismissals vanquished national, intermediate, and local representative bodies. Generally, when a representative was dismissed, the representative from the lower house would replace him. But with so many representatives dismissed, the last official standing for some spots, at the end of the line, was a rural teacher who had held a local post. Most were, fortuitously, members of the prime minister's party or of a coalition party. In the prime minister's zeal to remove his opposition, he actually elevated a few select Ethiopian women—a small step for Ethiopian womankind achieved purely out of coincidence.

Once the ladies finally became comfortable, and after some prodding on the translator's and my part, they began to perk up. The first one spoke, looking less at me and more into the open space straight ahead of her. She talked about how impressed she was that I was a lawyer and how she never thought she would meet an American Muslim female lawyer. Another woman made similar comments (all translated) with a similar demeanor. Soon all the women were speaking in turn, still not quite smiling but jubilant that I was a lawyer. I told the lady parliamentarians how greatly I admired their election to office. In my eyes, their achievements were greater than going to law school. Again, they returned the translation with blank stares. Their reactions, their inability to open up, were bizarre to me. One of them even said that what they did was not anything special, nothing to brag about.

I wanted to take a picture with the lady parliamentarians, to remember my visit with Hillary's and Nancy's Ethiopian counterparts. Upon exiting Parliament, where my camera had been held at the door, I started signaling for the ladies to gather around the entrance to the government building, a natural backdrop for our photo. The ladies were hesitant, and then many of them started lining up to go for it after all.

Immediately, one of the two guards posted at the door rudely chided the ladies. He gruffly told them to move and insisted that photos could not be taken there. I didn't realize what was happening, but I did notice the ladies acquiescing quietly and moving away. I followed the ladies and asked what happened. The translator told me that the guard had chastised them for attempting to take the photo in front of the building. Why, I asked the translator, would a guard tell a Member of Parliament what to do? That would be like the metal-detector guard at the U.S. Congress telling Hillary she couldn't take a photo in front of the U.S. Capitol. It would be an outrage.

The ladies quietly moved on, taking their chastisement without so much as a frown. Their stares were the same as I saw in middle-aged women all over Ethiopia. We took our photo in another location, and the ladies dispersed without even so much as a good-bye. Ethiopian women, even Members of Parliament, don't linger. When these moments of injustice take place, I always think I will stand up against them. But, as with many things, the individual events happen so quickly that it's over before I can protest. I did nothing.

More than a year after I met them, I thought of the Ethiopian women while I was standing outside the Garden Grove Mosque in Southern California. My uncle Adnan was telling me how my grandmother's burial service would proceed: "The imam and the mosque people want the women to stand in the back, because women get emotional." I gave him a blank stare back, which is an unusually passive response for me. He added, "Women start crying, and then they might throw themselves or fall in the grave." I had to resist telling him that we were not in a Bollywood movie from one hundred years ago and that I had enough sense not to jump into the grave.

Visiting Ethiopia a year earlier had made me feel lucky to live in America, where varying and misguided religious interpretations do not often affect me directly. But on this day of grief, I was confronted again with the Muslim community's extra-Qur'anic patriarchy. It rears its ugly head, and it is the rare woman who has enough education, confidence, and support to say *no, this is not what Islam says; this practice is a cultural one.* I didn't feel like arguing the point with him. I just said quietly and apologetically, because I did sympathize with how the mosque folk would hold him responsible for my insubordination: "I'm sorry. I can't do that. I am going to stand wherever I want."

She was my grandmother, after all. I had every right to watch her go into the next world and a duty to be a witness for this event. She would have done the same herself, I imagined. She possessed a bold personality, unlike Western stereotypes of Muslim women and unlike the lower-level status that Muslim leadership forces Muslim women into today at the mosque, in organizations, and socially. She was headstrong and lucky that she moved to the United States in the late 1970s, away from Pakistan when Wahhabism and conservative views began gaining in popularity. My grandmother was full of life and emotion. She quite infamously, upon meeting a suitor for my sister's arranged-marriage years ago, hugged and kissed him and gave him a one-hundred-dollar bill, per the Pakistani tradition. (Despite this bribe, surprisingly, it didn't work out for him and my sister.) In fact, for years she even gave my father a one-hundred-dollar bill whenever she saw him and, before she became very sick, would FedEx a one-hundred- or fifty-dollar bill to each of her grandchildren for our birthdays. At some point, I was downgraded to fifty dollars. I am not sure why, but maybe it had something to do with my not hiding my true age when I turned thirty-one. (My Naunni had a problem with this because, according to her, how could I be thirty-one when she was only twenty-seven herself!)

My grandmother was reassuring. When I was sick, she would always say, "*Aaap mere sath ro-ho.*" Literally, it means: "You live with me," but the phrase is actually more special than its translation suggests. *Mere sath,* literally "with me," means more. Although one can be with another in physical presence or on one's side in a debate or argument, *mere sath* means both a physical and emotional togetherness. Naunni was a big proponent of the *mere sath* because she truly believed that if you ro-ho (or lived) with her (*mera sath*), you could be in her force field, her geographical energy field, and that she could heal you of anything that ailed you. She always made an effort to see me when I was sick, or at least talk to me on the phone. She had great self-confidence in everything—in her looks, her adamant belief that her age was indeed only twenty-seven, and her abilities to heal just by her mere presence. I tried to heal her in return during her own illness—to heal her with my presence—but it was not to be.

Most of my memories of Naunni are of when she was more herself, before cancer robbed her of vitality. She was so young looking—not a wrinkle, jet-black hair, skin that glowed with color. After her death, her pearl-like color returned.

My mom wrapped one of my grandmother's embroidered scarves around her head. She looked like an otherworldly, ethnic Madonna. She would wrap her scarf around her occasionally when she was alive, too, but out of a sincerity of feeling—when she was touched or moved by emotion or feeling a little anxious—never out of the Wahhabi compulsion that necessitates it today. She wasn't Wahhabi, but she was a victim of patriarchy in a different way. She insisted on living with Adnan, her youngest son. I think she believed that, as in Pakistan, if you lived with your daughter and son-in-law, you could be kicked out at any time. Rumor has it that my great-grandmother was ejected from her daughter's home (the sister of my grandmother), on the order of her husband, back in Pakistan. Another sister of my grandmother's, who couldn't afford to keep her at home, took her to a hospital in her home village where she died soon after. My grandmother claims that the moment she found out that her mother was in a hospital, alone and abandoned, she set off to see her, taking a train and then a horse-drawn buggy. She used to say that when she entered the hospital room, she felt her mother take her last breath at that exact moment and die. I think my grandmother took from that experience to always live with your son, because he can order his wife to keep you.

Muslim women always have to navigate between patriarchy and making their own decisions. Many Muslim men seem to believe that individuality doesn't exist in Islam, especially not for women. These men force this paradigm on women. The War on Terror, under which even nonterrorist Muslim men feel threatened, reinforces this dichotomy. For a woman to choose to be herself is interpreted as demolishing the faith—a victory for the West, which is perceived as wanting to destroy Islam. When people like me can't speak out, the cycle is perpetuated. The consequences of my not speaking out could be dire. I feel that, in this environment, it is important to remind Muslims why we are Muslim and to tell non-Muslims what Islam really stands for. My hope is that Muslims will stop forcing emotionally destructive interpretations and that the West will reach out to us with more understanding. I don't want any more blank stares on either side.

"You were being friendly with the imam," my sister jocularly accused me after my grandmother's burial. I am quite anti-imam, so for me to say hello, as I did just before the burial, was surprising. By saying hello, I meant to put him on notice that I was there and to tell him that many women who loved Naunni were there. He had praised my writing in the past and remembered me when I reintroduced myself. After the men had thrown a symbolic fistful of dirt on the grave, the imam called the women over. Like the Ethiopian lady parliamentarians taking a photograph outside Parliament, they were hesitant at first but slowly came over. He even called in the women who sat outside the graveyard fence. The imam made sure each woman who wanted to release a handful of dirt did so. Some women never moved past the fence, too reluctant to break with the patriarchal practice. But the imam and many women broke with this non-Islamic tradition. I hope that other Muslim leaders will, like this imam, see that women have a role to play too.

I hadn't had the energy to organize a social protest, to gather all the women and force the men out of the front spots. But I didn't agree to stand in the back either. Adnan had cited his sister (my aunt) Tofi (who is branded as "emotional") as an example of why women must stand in the back, and I told him, "You can tell her to stand in the back, but I don't have that same problem." I would be like the Ethiopian women, unemotional, staring ahead, if that's what it took not to be relegated to the back. I understand the compromise the Ethiopian women made. At my grandmother's burial, I did the same thing. I was not strong enough on that day to change the system completely. But I could do my part, and I will keep speaking up and making my own small protests. To tell the story of why I am a Muslim is to speak up. It is a small protest against the current waves to remind Muslims—and to make non-Muslims aware—of the true, motivating values of Islam.

—A.G.H., 2008

red, white, and muslim

Chapter 1

Born Muslim

He it is Who shapes you in the wombs as He wills.
There is no deity save Him, the Almighty, the Truly Wise.

Qur'an 3:6

"When forty-two nights have passed over the drop,
God sends an angel to it, who shapes it and makes it ears, eyes, skin, flesh and bones.
Then he says, 'O Lord, is it male or female?'
And your Lord decides what he wishes."

Hadith (saying)
attributed to the Prophet Muhammad,
from the Hadith collection of Sahih Muslim

I grew up in a rural, minimetropolis about one hundred miles south of Denver: Pueblo, Colorado. Our house was Mediterranean-style stucco with a red-tile roof. We were the sole Muslim and Pakistani family for years and, even with a few additions, the Muslim and Pakistani population was never high. Pueblo is the gateway to the southern Colorado community and the Southwest. It had a large Latino and Chicano population, of which I became an honorary member because of my dark hair, eyes, and complexion. Native Spanish speakers would solicit me in conversation.

"*Como está?*" an older Latino man would say to my sister and me while we waited for my mom to pay for her groceries.

"Oh, we don't speak Spanish," my sister would say authoritatively, taking her role as the older sister very seriously.

"Bah, you kids don't care about your culture no more," the man would gruff and then stomp out of the store. Spanish speakers would become very irate when we wouldn't respond "*en español.*" But as we would later tell Mom at home, we weren't Latino.

"Yes, you are!" My mother replied.

"Mom, how can we be?" my sister asserted the way that an eleven-year-old girl does with her mother.

"Your ancestors were the Moors," my mom said, "who conquered Spain."

"But Mom, I thought we were Mongolian," I whined, confused.

"Your mother claims to be related to everyone!" my father's phantom voice piped in from the background.

As my mother had told me many times, she once again related our history: In the late thirteenth century, the descendants of the Mongolian warrior Genghis Khan traveled just off the Silk Road from their Central Asian home to an area in what is now called Afghanistan. The grass was high. Tigers and lions roamed freely then. The Mongolians wore long, parka-like coats with fur collars and cuffs to stay warm in the winters. They had fair skin but dark hair and Oriental-looking eyes that were narrow and smooth. They were Muslims. They were the Khans. Their people would live in the shadow of the Himalayas for centuries.

Hundreds of years later, in 1970, Seeme Khan, who belonged to the family branch that had moved to India and then Pakistan, married a man from a different ancestry—the Aryan tribe of India, descended from migratory Europeans—who took her to America, where she gave birth to me in Chicago. I was born a Muslim and on the small of my baby back was the blue blemish all Mongolian babies are born with and which eventually fades away in infancy. Called the Mongolian spot, it is found among direct descendants of Genghis Khan.

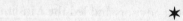

✶

My mother wanted to name me Scheherazade after the narrator-character of *The Arabian Nights.* My grandmother wanted to name me Asma (ahh-si-muh), after her sister who died young and whom she never knew. My mother found the choice a little macabre and said, "We'll let her father decide." When my father came to visit me at the hospital the day after I was born, he said, "I love my mother-in-law so much, she can have whatever she wants." So my father, much to my mother's shock, named me after her late aunt.

Asma means "high" or "exalted" in Arabic. *Asma* is derived from the Arabic and Persian word *asmaan,* which means "sky." Although my dad is a doctor, he didn't realize that people would forever be calling me "Asthma." Throughout my school days and until only recently, boys would mockingly breathe heavily in front of me, simulating an asthma attack, each one thinking he was being quite original. The latest assault on my name comes courtesy of Microsoft Word's AutoCorrect feature, which automatically corrects my name, ASMA, to ASTHMA. Once a law-school professor of mine wrote an e-mail to me reading, "Dear Asthma, I know your name is Asthma and not Asthma, but my Microsoft Outlook e-mail is automatically changing Asthma to Asthma. Sorry!"

My great-aunt Asma, and as a result I, had been named after one of Islam's bravest women: Asma bint Abu Bakr (which roughly translates as "Asma born of Abu Bakr"). Abu Bakr was one of the Prophet Muhammad's closest advisors and led the Muslim community after the Prophet's death. Asma was one of the first converts to Islam. In 622, Muhammad heard news of an assassination plot against him. His fellow Meccans, fed up with his talk of this new and just religion, were going to finish him off before he could gain more converts. Unlike the rest of the early Muslim community, Asma had not yet escaped Mecca for Medina. The Muslim community had been invited to resettle there by the locals in exchange for Muhammad's services as an arbitrator between Medina's factions. Now Muhammad was going to make the journey to Medina, not just to resettle but to flee from the attempts on his life.

In the dark, desert night, and with his cousin Ali sleeping in his bed, Muhammad sneaked out of Mecca, en route to

Medina. Abu Bakr accompanied him. They took no provisions. Just in case someone did recognize them in the dark, they didn't want them to realize that Muhammad was in the midst of an escape. They knew they would be camping out in the Arabian desert for a few days to let the murder plot unravel. They hid in a cave with a small opening.

The next day, Asma set out for the cave, a spot that had been chosen beforehand. She had bags of food and water hidden on her body, which she was going to sneak to her father and Muhammad. She had to be extremely careful. Everyone was looking for Muhammad now as the deception of Ali sleeping in his bed was revealed. A bounty was put on Muhammad's head: anyone bringing him to the ruling elite of Mecca dead or alive would be handsomely rewarded. But Muhammad might have ended up dying of starvation and thirst first. He would dry up in the desert cave before any bounty hunter could find him. Asma made sure he did not.

For several days, she brought them food surreptitiously, and not one of the ruthless Meccans found Muhammad. She was pregnant at the time, too, and soon she escaped to Medina where she gave birth to her son Abdullah just outside the city limits. The early Muslims considered Abdullah's birth a blessing. He was the first Muslim to be born to the now-free Medina Muslim community. Asma had lived up to the meaning of her name: sky. Like the sky, she sheltered Muhammad and her father, securing Islam's growth and success for the next twelve years, when her father's leadership of the community ended with his death in 634.

*

When I was a baby, my grandmother told me the story of Asma bint Abu Bakr and her own sister Asma. She spoke to me in Urdu, the language of her home country, Pakistan, while I dozed in and out of naps as a baby does. When I was a child, my parents spoke to me in the tongue of their youth: British-Pakistani English. But when they wanted to tell each other secrets, they spoke in Urdu, not realizing that I had already dreamt in that language as a baby. When my mom talked to her siblings on the phone, she sometimes spoke in Punjabi, a secondary language of Pakistan. I spoke American English, the language taught in school.

With all these languages in my head, I guess I could not keep them straight. In first grade, my teacher told my mother that I was "retarded." My mental disability, she said, meant I would never learn to read or write English without special help. I sat quietly watching my teacher tell my mother that I was retarded. I had a disease, my first-grade self thought. I accepted it. I knew that all the other kids in the class knew the alphabet better than I. It took a lot longer for them to say it than for me: A B C K L V Z was my alphabet, roughly.

My mother was defiant at my diagnosis. She said to the teacher: "My daughter *can read*. You just don't know how to teach her."

That day, before we had dinner, my mom called me downstairs to the playroom in our basement. "Sit down here," as she pointed to a spot next to her on the couch. She had a book of the alphabet in her lap. Thus, I began learning British English from my mother, every day, after school for an infinitely long hour. English had, until then, been a haze to me, a blur of black and white letters and pages, out of focus and flashing by me.

"Make the sounds," my mother would tell me. "C sounds like cuh, cuh."

"Cuh," I said breathily one afternoon, "ah-tuh. Cuh-ah-tuh." Recognition shook my entire little body. I knew what this was!

"CAT!" I squealed proudly. The blur of the letters had started to come into focus just a little bit. I was so excited.

"That's right," said my mom.

When the Prophet Muhammad received the first Qur'anic revelation, he had been meditating alone in a cave, as was common in that time. He was a religious man without a religion. He had heard of Christians and Jews and wondered why his people did not have a movement like those. Out of nowhere, an apparition appeared before him—it was not a person but it looked like one. The apparition squeezed him, hard. Muhammad felt like he couldn't breathe.

"READ!" the apparition commanded him. It was the angel Gabriel. The Arabic term for "read" is *Iqra.*

Muhammad, now practically suffocating, knew very well he couldn't read. He was illiterate. Having been born an orphan into a poor family, he was not educated.

I can't! he must have thought, pleadingly.

The angel insisted again, "READ!" IQRA!

Muhammad wasn't sure if he was going mad or not, but he felt the angel right there in front of him, squeezing the life out of him.

"READ in the name of thy Lord who created you," Gabriel insisted.

Caught in this furious embrace, he felt the words rise up out of him. Something had taken hold of him, and the angel's grip kept him from fighting it. His mouth began moving with the most amazing words:

> *Read in the name of thy Sustainer, who has created—*
> *created man out of a germ-cell!*
> *Read—for thy Sustainer is the Most Bountiful One*
> *who has taught [man] the use of the pen—*
> *taught man what he did not know!*
>
> Qur'an 96:1–5

Muhammad received the first revelation of the Qur'an. Islam was born. It would change the life of Muhammad, his wife Khadijah, and, eventually the lives of more than a billion people, including me.

Muhammad, despite receiving revelations till his death twenty-two years later, never learned to read. Reading was a luxury in pre-Islamic Arabia. In fact, even today most of the Islamic world is illiterate. Unfortunately, the Qur'an that Gabriel exhorted Muhammad to READ cannot be read by most of the world's Muslims. I would not have this problem. I was lucky enough to be born to a mother who could read and who taught me British English. The homework my mother sent me off to school with would read: "colour" instead of "color," "theatre" instead of "theater." I would return home with the graded paper, which my teacher had marked as incorrect with these spellings.

"No wonder she thought you couldn't read," my mother said. "She can't even spell!"

By the end of the year, I was the strongest reader in my class—in any kind of English. Not bad for a "retarded" Mongolian.

I was born a Muslim. When you are born Muslim, you are born with the history of Islam behind you. The birth of each Muslim continues the epics of Islamic history. My Islamic ancestors were great scholars, conquerors, and philosophers. They invented algebra and modern navigation. They fought the Crusades, and they brought peace to Arabia, Spain, and India. Islam is more than a religion. Islam has been one of the world's greatest and most successful movements ever. The famous Pakistani poet Iqbal wrote of the cycle of Islamic history that "Islam is revitalized after every Karbala." By referencing the Karbala massacre, where Muslim factions fought each other, killing the Prophet's grandson Husain in the process, Iqbal reminds us that the story of Islam is one of both disappointments and triumphs. The poem continues by saying that for every time, for every group of Muslims, they must face their own version of a Karbala before they can move into the next phase. Just like the Moorish Muslims spoke Spanish in Cordoba, this South Asian Muslim girl spoke Spanish in Pueblo.

Islam teaches that we are all born Muslim, actually—a concept called *fitra*. Then, over time, we lose touch with the religion of our birth. Many are assimilated into another religious tradition, either because the family the child is born into is not Muslim or the child's surroundings and environment don't facilitate spiritual growth. *Fitra* should not be misunderstood though. When Islam says that we are all born Muslim, Islam is saying that we are born wanting to submit to God's will, that we are all born innocent and able to recognize right from wrong. The word *Muslim* means a person who follows Islam.

Islam means submission. The implied object of this submission is God. A Muslim is one who submits to God and God alone. *Fitra* is the idea that we are born ready and able to submit, inherently capable of doing right instead of wrong:

> *And so, set thy face steadfastly towards the one ever-true faith, turning away from all that is false, in accordance with the natural disposition [*fitra *in Arabic] which God has instilled into man: [for,] not to allow any changes to corrupt what God has thus created—this is the [purpose of the one] ever-true faith; but most people know it not.*
>
> (30:30)

Muslims believe that God, as the Creator, created all humans and that He meant for us to be diverse. Our diversity is also a part of our collective *fitra*. As a Muslim, I am part of a world community of over a billion people. The Qur'an says:

> *And among His wonders is the creation of the heavens and the earth, and the diversity of your tongues and your colors: for in this, behold, there are messages indeed for all who are possessed of knowledge.*
>
> (30:22)

A Muslim meets other Muslims almost daily. A baggage handler at Denver International Airport my mother recently met told her about his escape from Ethiopia. Like Muhammad and Abu Bakr, he escaped Ethiopia in the middle of the night, fleeing to a refugee camp. Once he received refugee status, he was granted asylum in the United States. When they met and realized the other was Muslim, they gave each other the Islamic greeting that all Muslims are taught, from when they can first talk, to give each other. Even Busta Rhymes—the hip-hop artist known for his animated presence—returned my Islamic greeting when I saw him once at a hotel.

"*Assalamulaykum,*" is the greeting Muslims give to each other: "Peace be with you." "*Walaykumasalaam*" is the response: "And also with you." From Busta to the DIA baggage handler, the fitra manifests itself. When I meet another Muslim, I automatically know I am meeting someone who understands my heritage, who believes in one God. This recognition between Muslims is part of the fitra of all Muslims. We are over a billion brothers and sisters, living all over the world and yet sharing core beliefs. At any moment, somewhere in the world, a Muslim is performing one of the five daily Islamic prayers. As Muslims, we take pride and comfort from this unity and acknowledge it with our special greeting when we see each other. We understand each other at a level so intimate it could almost be subconscious. This connection is spiritually powerful.

When God made Adam and Eve (called *Hawaa* in Islamic tradition), He made them independently of each other. He made them because He wanted one of His creations to represent Him, to exemplify the knowledge He had but that His other creations—the angels and animals—lacked. When they both used their curiosity to eat the fruit of the forbidden tree, God had to punish them. He expelled them from Eden to Earth where according to Islamic legend they were separated from each other. Although God forgave them, He wanted them, in Islam, to find each other first.

For years, according to the legend (which is described in Kanan Makiya's lyrical novel *The Rock*), Adam and Eve wandered Earth alone. They were constantly moving and searching but not knowing what they would find or even what they were looking for. Their memories of Eden were distant. When they finally ran into each other at Mount Arafat in today's Mecca, the recognition was immediate. Remembering their idyllic past, Adam must have asked, "Is that you?"

"Who else could I be?" Eve might have said in response. It was *fitra* for them to recognize each other—they were predisposed to be the other's mate. God had made them that way. It is *fitra* for me to be a Muslim. When Adam first laid his eyes on Eve, his connection to her was intuitive. For Muslims, the connection to God is intuitive—like when Muhammad finally began to read as Gabriel squeezed the breath out of him or when I first realized I could read the word *cat*. I was born to be Muslim and made by God to be a Muslim.

Every day in grade school at John Neumann Catholic School, when my classmates would have religion class, I would go to the principal's office. Between his office and the large administrative office was a small room, almost like a hallway, where students would sit when they were in trouble and waiting to see the principal. (Or, in my case, when you were excused from a class.) The room had a few wooden desks, the kind where the writing area comes out from the side like the chair has an arm. They were hard, and different students had chiseled their names or a phrase into the dark brown wood.

I WAS HERE, one marking said. So was Jesus. A small likeness of him hung above the doorway on a small crucifix. I would drag my fingers along the grooves of the desk face, feeling the gaps and indents where students had marked their territory. I was sent here to read books on Islam my mother had bought in Pakistan. She had struck a deal with the school. I did not have to attend the religion class as long as I used the time to study my own religion. Although, by that time, the nuns had made sure I knew all the Catholic prayers. I liked the Apostles' Creed the best. It had a lyrical quality:

I believe in God, the Father almighty…
I believe in Jesus Christ…
He suffered under Pontius Pilate,
Was crucified, died, and was buried…
He will come again to judge the living and the dead.

Once, the principal was en route to his own office. Walking through the little room, he caught me reading an *Archie* comic book. He had longish brown hair and a beard, and I always thought he purposely tried to look like Jesus with his beard and hair. He even told us once he had played Jesus in a church event enacting the Stations of the Cross.

He picked up the Islam book I had brought with me but was ignoring. He opened it to the bookmark and set it on the desk in front of me. Holding the book open with his hand, he stood in front of my seated self.

"READ," he commanded. He then marched off into his office. He was normally a very gentle person. This was the most aggressive I had ever seen him.

I looked at the page he had opened to. It was headed "The Asma al Husna." What was this, I said to myself. Is this my name? I had never seen it in print before! I was so excited. My name was here on the page in front of me: A-S-M-A. I immediately started reading. God is described in the Qur'an as having many attributes, like mercy and power. God has several names—each one describing one of these attributes—ninety-nine names in all. The concept of the Asma al Husna is that God is present everywhere and in all ways. The Asma al Husna is a way for us, as Muslims, to acknowledge this belief. Some Muslims recite the various names using prayer beads throughout the day. It's not uncommon to see a Muslim sitting in a corner, holding a chain of beads and fingering them one by one, quietly saying each name to himself or herself.

The Qur'an says, "The most beautiful names belong to God: so call on Him by them" (7:180). One of the names is Al Khaliq, the Creator, for God created us. He created us as Muslims. The Asma al Husna is so called because *Asma* means high or elevated. These ninety-nine names are the names of the God who is the most high and most elevated of all: the Asma al Husna. I was so thrilled at this find. My name—which had been the source of jokes mainly until then—really stood for something neat, something very special.

I told my aunt Tofi about the Asma al Husna that weekend when we visited her in Aurora, a suburb of Denver, where she lived. I told her like she had never heard of it before, which, of course, she had. "Can you believe all those names are called Asma?" I asked exuberantly.

"Well, that's a better title than 'Asma the Great,'" she said, referencing the nickname I had made up for myself. (I had been learning about Alexander the Great in school when I came up with it. I figured if Alexander had the title, why couldn't I?)

"Oh yeah," I said. "Well, I don't think I'll change it," I laughed. "Asma al Husna is for God!" I chided her.

Years later, my aunt Shazi told her husband, my uncle Imran, "I want to name our baby Sara, after Abraham's wife."

"Okay," my uncle said in his accommodating way.

"Why Sara?" everyone asked my aunt. We were Muslims, and why wouldn't her child have an Arabic name like the rest of us?

"The name Sara is a part of Islamic tradition," Shazi told me when I saw her after the baby's birth. "And I don't want Sara to have all the problems you have had with your name." I suddenly regretted all the complaints I had made about my name, how no one pronounces it right, how everyone makes fun of it, how awkward it is spelling it all the time. Even though Sara is as much an Islamic name as Asma, I felt bad that I had made the name Asma seem like such a burden. It is a burden, but it is also a blessing. It is who I am. I would never change it.

"Well, good. She won't be harassed all the time like Asma was," my sister Aliya blurted out, joining our conversation. I looked at the baby—she had fair, almost pink skin and a wealth of black hair sprouting from her head. With a few exceptions all the babies in our family tend to look identical at that age; they're always large and healthy with the Mongolian spot. The few exceptions, like me, are born premature and small as a result. I forgot to ask Shazi if Sara had the blue blemish on the small of her back like Aliya and I did.

"She has eyes like my sister's," my grandmother said. Her sister, Pino, is Shazi's mother. But Shazi did not inherit Pino's special eyes. Sara and her maternal grandmother have what my family calls the "Mongolian eyes," with narrow eyelids and a smooth space under each eyebrow. The eyes are another vestige of the long line of Islamic people we come from—another aspect that ties us to a history, a tradition, a global community, and a way of life.

Chapter 2

A Direct Relationship with God

It is We who have created man, and We know what his innermost self whispers within him: for We are closer to him than his neck-vein.

Qur'an 50:16

I was trying to study for my spring-semester law-school exams and hadn't been doing too well when the phone rang. Outwardly, I had been blaming my lackluster study habits on *Ramadan,* the month on the Islamic calendar during which Muslims are required to fast from just before sunrise to sunset. I had been fasting almost the entire month. But to be honest, the fasting didn't affect my studying too much; I was just being lazy.

My sister's friend Monir was on the line. He was calling to ask if I wanted to attend Eid prayers with him at the Islamic Cultural Center of New York near his apartment—the large famous mosque on 72nd Street on the Upper East Side of Manhattan. The Islamic holiday that comes at the end of *Ramadan,* Eid al-Fitr, was coming up. Muslims gather at their local mosque to pray together in celebration of the holiday. After attending a reception at the mosque, most Muslims have a lavish feast at home and donate food to charity that day.

I arranged to go with him, and the next morning I put on my newest *shalwar-kameez* (Pakistani ethnic dress, consisting of loose-fitting harem-type pants and a long tunic) and flagged down a cab on Second Avenue for the journey uptown. The driver took the FDR Drive, and we whizzed by the dark East River, which held court over a promising but cloudy sky. The mosque actually had three different prayer times to accommodate all the Muslims who would come. For some reason, everyone, including Monir, always wants to go to the first one. Normally, such an early morning would be a problem for me, but I had been waking up at 3 a.m. for the last month to eat a pre-dawn breakfast. So 8 a.m. practically seemed like midday. During *Ramadan,* by the time I attended classes around 11 a.m., I was ready for bedtime! *Ramadan* is very challenging—not just because you cannot eat but also because, if you are in a non-Islamic country, you don't have the luxury of staying home and sleeping in. (Those of us in law school are forced to sleep in class instead!) When Eid finally comes, the excitement of a holiday is joined by the excitement of returning to a normal daily routine.

We walked over to the mosque together from our meeting place—the apartment of another friend of Monir's. Other New Yorkers slowly started to dot the sidewalks as the early morning yielded to the beginning of the workday. Aside from my ethnic clothing, Monir, his other friends, and I could have been any other New Yorkers—young, twenty-something, professional types.

As we came closer to the mosque, I saw more and more people—American Muslims in all different types of clothing and all different colors. We had only walked a few blocks, but we went from desolate New York morning streets to the flooded perimeter of the mosque. Hundreds of people were milling about. Monir and one of his male friends headed off to the men's entrance, and I went with Monir's female friend to the women's. Normally, I won't even go to a mosque that has separate entrances for men and women, but I make exceptions on holidays. In any case, the crowd was so large that, here, separate entrances managed crowd flow.

The Muslim prayer area is supposed to remain clean, so anyone wanting to enter must take off his or her shoes. Like most mosques, this one had rows and rows of cubbies for people's shoes. By the time we arrived, most of the cubby space had long since been filled up. Already, hundreds of pairs of women's shoes were piled on the floor.

Peeking inside the prayer area, I saw all these women gathered, most sitting on their knees, shoeless, and awaiting the beginning of prayers. A little girl was wearing a tiny piece of fabric stitched into a sort of instant head cover. She just slipped it on, and she was ready to go. In a few moments all of our voices would be, in unison and then individually, praying to God. For every pair of shoes lying here was a woman inside who was about to pray—a woman with hopes, longings, anxieties, grievances. And God was about to hear us all. I pulled my scarf over my head and entered.

The prayers began soon thereafter. As a group, we did the motions of prayer in unison—kneeling, standing, leaning over, standing again, and so on. We put our foreheads to the carpeted floor with our hands at our sides, palms also against the carpet in a position of absolute vulnerability and submission. What we are saying with this pose is, "I submit to You, God." Even with all these women submitting en masse, I knew that God heard each one of our voices individually.

In the Qur'an, God is often described as "all knowing" and "perfect in knowledge." God knows all thoughts that enter your mind—your prayers, your desires, your sadnesses, even those moments when you are thinking about having a candy bar or wondering if you remembered to lock your door on your way out. When you hear this aspect of God, your first reaction is to say, "Impossible—to know the thoughts of all people!" But we forget that God is not a person like us but a divine being. We have no idea how powerful God is. The ability to hear the prayers of the hundreds of women I was praying with, and mine too, is a God quality, not a human one.

That God knows what you are thinking at all times is, of course, something of a mixed blessing, but mostly I am emboldened knowing that every thought I have, everything I do, is a prayer to God, whether I am in the mosque or not. I do not need to attend a mosque to talk to God. I have my time with God whenever I want. When I do pray, I pray directly to God. In Islam, God has no partners or intermediaries. He alone is the final judge of us all. Islam has no middleman. God says in the Qur'an that we are each close to him and that he listens to us: "If My servants ask thee about Me—behold, I am near; I respond to the call of him who calls, whenever he calls unto Me: let them, then, respond unto Me, and believe in Me, so that they might follow the right way" (2:186). No parts are sold separately in Islam: God comes in each unit of life. I don't have to rely on others for access to God. Neither did Hajira (known as Hagar in the Old Testament), although she didn't realize it. When Hajira, mother to Abraham's first son, Ismail (Ishmael to Christians), ran between Safa and Marwa in Mecca, where Abraham had left her and his son to fend for themselves, looking for water, God knew what she was going through. He sent Gabriel to help her. Hajira did not know that she had a personal relationship with God. God looked out for her (as He does for all of us) the way a parent does—always available, at times a figure of discipline but mainly to support.

Islamic folklore tells the stories of a few individuals who lived before Islam existed—individuals who believed in one God but had not yet found the religion to go with their beliefs. Such an individual is called a hanif in the Qur'an, derived from the Arabic word *hanif,* which means "upright" and "moral." A *hanif* I once read a story about was so obsessed that he left home as a teenager and desperately searched through the Arabian desert on horseback for the message of the one God. I imagine his horse galloping furiously through mounds of desert sand and the heat. When he finally met Muhammad, he had all the frantic energy of someone who had lost something very valuable. "Are you the messenger?" he must have asked Muhammad. Was Islam the path to God, finally?

Islamic tradition believes that Abraham was such a *hanif.* He made a covenant with God to believe in Him—the one, true God. Abraham's covenant was the beginning of this direct relationship between God and the people who believed in Him. Muslims believe that many people were also seekers of the one God before Judaism and Christianity brought monotheism to their people. Islam brought monotheism to the people of Arabia and then, eventually, to my ancestors.

After my family would be Muslim for hundreds of years, far, far in the future from the *hanif* on horseback, I would find myself, an eighteen-year-old at the beginning of her senior year, facing another *hanif.* I was on an island off the coast of Massachusetts—the site of an Outward Bound course my boarding school had taken all the senior students to before the start of the school year. The breeze off the water was enough to counteract the early fall heat. Generally, the day would have been pleasant except for the fact that we were divided into small groups consisting of people who were not really our friends. I looked at my group. Except for one or two people, I was not friends with anyone in my group or was intimidated by them or both. The day was going to be rotten, officially.

"Your first task," the Outward Bound leader told us, "is a game of keep-away. You have to keep someone in your group from taking this!" He held up a piece of pastel-colored cloth tied up in a ball.

"Imagine that it's a rare and valuable jewel," he told us. *Could anything be more lame,* I thought. I would have rolled my eyes if I hadn't been so worried about the impression I was making on my teammates.

"Now, who's going to be the one to take it from the rest of you? Who has the best shot of stealing this from all of the rest of you protecting it?" We all immediately agreed—in our first and only show of teamwork that morning. The one person in our group who could somehow foil us was probably also the only one in the group I genuinely considered a friend. His name was Ghani, and the only reason I had ever spoken to him was because he was Muslim. He was one of the two other Muslims in my class and the first African American Muslim I had ever met. He was also totally wacky and had bionic gymnastic-type abilities—I think that's why he was nominated for the impossible job of stealing the jewel from the rest of the ragtag group.

Ghani was exiled while the rest of us, who had probably never said more than a paragraph's worth of words to each other, had to plan how to keep Super Ghani from stealing the jewel. We had to keep the jewel on the ground. None of us could hold it. In hindsight, our perfect plan showed how little we knew about each other. We decided to make a circle around the jewel, with our arms stretched out to the sides, and just improvise depending on what Ghani did.

The grass was such a bright green that with the sun, it was almost blinding daylight. Ghani was single-minded, like the *hanif.* He did a somersault in the air over our circle, landed long enough in the middle of us to grab the jewel, and then leaped over the lowest set of arms. We didn't even have a chance against him. We all looked as a group at him, standing about four feet from our defiant and useless circle.

"Hoooowaaahhhh!" Ghani howled as he made slow-motion kung fu gestures with his hands and legs. We did the same exercise a few more times, and each time Ghani prevailed. Why didn't we try something else besides our circle of doom? Why didn't one of us take Ghani on one-on-one?

Or why didn't several of us surround Ghani or start leaping around him? We were totally unfocused—trying to keep Ghani away from the jewel but still trying to protect the jewel. But Ghani was focused on the jewel. Like the hanif, galloping through the desert, he knew what he was looking for, and all he had to do was take it.

God created us to be good and innocent, pure and predisposed to good. We are human, but we are simple creatures too. The Qur'an describes faith as a "truth" that, once we have found it, should guide us like a needle on a compass would: "So, set thy face steadfastly towards the [one ever-true] faith, turning away from all that is false, in accordance with the natural disposition which God has instilled into man: [for,] not to allow any change to corrupt what God has thus created—this is the [purpose of the one] ever-true faith; but most people know it not" (30:30). As Hafiz, the famous Sufi poet, put it: "I have lit that lamp that needs no oil." God provides the continuous light. We, as humans, become caught up in the confusion of daily life and either are distracted from the light or abandon the search for it altogether. Other people try to confuse us, too, or are drawn to being mischievous. Instead of seeing the jewel, we see the people blocking our way, or we see the other people going after the jewel. Islam makes it simple: God is here for you. Will you be like the *hanif* and take God?

After lunch, our team finally began working together. We climbed over a twenty-foot wall together (it took us a while). We did a ropes course. By the end of the day, we were practically best friends. Unfortunately, the effect wore off by the time we graduated. The people in my group who I was unsure of, I was unsure of again by graduation. But for a few hours on the afternoon when I would have rather been doing something else, we found something, and we used it.

The most important thing in Islam is belief in one God. Muslims are required to believe in one God. The pre-Islamic, Arabian pagans believed in multiple gods. Islam's contribution was that God is one unified being. He is not multiple nor does he have partners. God has *tawhid* or unity. Just like God is unified, so the Muslim community should be unified too. *Tawhid,* the concept of God's oneness, spills over into other aspects of Muslim life. If God is one, then the community of Muslims should be one. For instance, Muhammad forbid the *ghazwas* or raids common in pre-Islamic Arabia. People would raid others' camps and caravans and loot their belongings. Muhammad outlawed *ghazwas* among Muslims. If God is one, then He should come first in all our lives. Focusing on what we can take from another person or what we didn't receive is not only destructive but also distracts us from God's oneness and unity. Unity—as opposed to disunity or discord—would and should prevail.

The hardest part about having a direct relationship with God is recognizing that the relationship exists. Not all of us are as astute as the *hanifs* were.

In Islam, prayer is the generally accepted way to communicate with God. A Muslim is directed by the Qur'an to cleanse himself or herself with water before praying—a ritual process called *wudu*. The idea is that you present yourself to God in your finest state. But if water is unavailable, that obstacle can be overcome. A Muslim can do *wudu* with sand. The *tayammum* (or the process of doing *wudu* without water) is a symbolic gesture to use pure earth to clean oneself if you cannot find water. So even the *hanif* who set out on horseback could still do *wudu* and pray if he had no water.

The Qur'an describes it: "And [you] can find no water—then take resort to pure dust, passing therewith lightly over your face and your hands. God does not want to impose any hardship on you, but wants to make you pure, and to bestow upon you the full measure of His blessings, so that you might have cause to be grateful" (5:6). The provision for using sand is God's way of telling Muslims that He is totally accessible. The absence of a component or factor of prayer does not mean God is absent too.

Islam has no formal requirement for a priest. Having such a requirement would counteract one of the foundations of the Qur'an—the idea that all people are equal. To place a minister between God and the worshiper would put up exactly the kind of barrier that Islam was meant to eradicate. Religion was not the place of the tribal leader, the local official, or the town minister. One's spirit belonged to oneself and to God—like a "Render unto Caesar what is Caesar's, and render unto God what is God's" on a massive scale. The Qur'an and Muslims are so dedicated to this aspect that, even today, no formal clergy exists that all Muslims are required to follow.*

At most mosques, the prayer leader also generally rotates simply to emphasize that no one person is in charge of the community spiritually. At some mosques small children, or even visitors, have led the prayer. While a mosque always has an *imam* who is knowledgeable about Islam and maintains the upkeep of the mosque, the *imam* is a leader in the operation of the mosque but not in your own spirituality. No actions of the *imam* or *fatwas* from any Muslim are binding on a Muslim.

Not only is no one person in charge of the community spiritually, but no one interpretation of the Qur'an is fixed

---

* About 10 percent of the world and the American Muslim community are Shi'ite Muslims. Shi'ite Muslims do have a formal clergy system. They believe that they should have a leader who guides their community spiritually and sometimes politically. The two major groups of Shi'ite Muslims are Twelver Shi'ites, who mainly live in Iran and follow the Ayatollah of Iran, and Ismaili Muslims, who follow the Aga Khan. The Aga Khan resides currently in Switzerland and uses the dues he collects from the world Ismaili community to fund charitable projects all over the world, including free hospitals and libraries. The research I did for my thesis in college actually benefited greatly from a fund the Aga Khan had given to the Massachusetts Institute of Technology (MIT) for books on Islam and Islamic architecture. My college had a library-loan program with MIT, which gave me access to books that I would not have been able to see otherwise.

for all time. This view is reinforced throughout the Qur'an. My favorite passage admonishes readers never to stop seeking the Qur'an's guidance: "Will they not, then ponder again and again over this Qur'an? Or are there locks upon their hearts?" (47:24). The Qur'an is meant to evolve and to be reapplied to each person. To judge how another practices Islam is wrong because God is the only judge.

The Qur'an may be old, but the interpretation and application to modern times can be surprisingly current. In fact, both Shi'ite Muslims and Sunni Muslims have incorporated this principle into Islamic theology. They believe in *ijtihad,* which roughly translates into Arabic as "free thinking." The idea is that a person of enough scholarship and training can read the Qur'an and reinterpret it based on the current situation he or she is facing. (All Sunni Muslims, who are the majority of Muslims, used to practice *ijtihad* until some Sunni Muslim scholars declared the "gates of *ijtihad*" closed.) All Sunni and Shi'ite Muslim clerics follow the other major sources of Islamic interpretive tradition—consensus, analogy, and consultation. For centuries, Muslim clerics have analogized the question facing them to a situation in the Qur'an or a reliable *hadith*★ of the Prophet Muhammad's to reach a conclusion.

The concept of *ijma* or consensus requires that the entire Muslim community agree on a principle or interpretation. Once this consensus has been achieved, the interpretation

★ *Hadiths* are statements the Prophet Muhammad is reported to have said and are often used by Muslims to evaluate a gray area in interpretation.

becomes an official part of Islam until the consensus changes. These interpretive approaches have been used by Muslims for hundreds of years and have kept the practice of Islam current.

In the post-9/11, pseudointellectual hubbub, a lot of non-Muslims buzzed that Islam does not believe in separating church and state, that Islam threatens our values. The truth is that Islam is so strongly dedicated to separating church and state that no Muslim feels the strong desire to constantly promote the separation.

Most Muslims, from America and elsewhere, belong to the Hanafi school of interpretation. This school, founded by Abu Hanifah, Nu'man ibn Thabit, emphasizes that the Qur'an is a living document. Furthermore, as Islam does not require any churches, mosques, or religious organizations, the separation of church and state is a nonissue. Islam may be the first widely followed religion to have such a strong stance. Non-Muslims don't realize that each Muslim, as a matter of Islam, has a personal relationship with God. So non-Muslims assume that the government must be speaking on God's behalf. Even in Islamic countries with a religious regime, the Muslims there feel personally connected to God. In Islam, you deal directly with God, and He deals directly with you. With each prayer, your sins are forgiven. You start with a clean slate. Islam is interactive.

The Qur'an describes God as powerful. God's abilities are expansive. God says He can see each one of us as an individual: "[For Him,] the creation of you all and the resurrection of you all is but like [the creation and resurrection of] a single soul: for, verily, God is all-hearing, all seeing" (31:28). Yusuf

Ali, the late Pakistani scholar, says of his popular translation in a footnote to this passage: "God's greatness and infinitude are such that He can create and cherish not only a whole mass, but each individual soul, and He can follow its history and doings until the final Judgment. This shows not only God's glory and omniscience and omnipotence: it also shows the value of each individual soul in His eyes, and lifts individual responsibility right up into relations with Him." I can choose to follow a religious cleric, a leader, an *imam*, or an *ayatollah* if I want to. Or I can keep communication directly between me and God, which is what the majority of Muslims do. God is able to minister to each of us personally.

With the freedom to worship God as an individual, though, comes responsibility. Every person, in Islam, is accountable for his or her deeds. Each person is required to follow Islam. "To whom much is given, much is expected," as they say. They must have been talking about Islam. The Qur'an says: "Your Sustainer is fully aware of what is in your hearts. If you are righteous, [He will forgive you your errors]: for, behold, He is much-forgiving to those who turn unto Him again and again" (17:25).

As Muslims, we cannot rely on others for our spirituality. We are responsible for our own actions. As the Qur'an makes clear, though, we will only be judged based on the spiritual guidance we have received—whether it is through Christianity or Islam or another religion or no religion. Although Muslims do believe that Muhammad's message is universal, if no messenger had been sent to your people, you would not be held to as high a standard as a person whose community had been touched by God's message.

While Muslims are freed from the tyranny of a supreme religious cleric, this freedom also means they are less equipped to counter criticisms and generalizations. No one Muslim can speak for all Muslims. So, when a tragedy like the attacks of 9/11 occurs, Muslims do not have a pope who can announce to non-Muslims that, as a group, all the Muslim community condemns 9/11. While all the Muslims I know condemned 9/11, non-Muslims needed to hear one spokesperson they could point to as the source of condemnation. The de-centralized nature of Islam, which is so beneficial to me as an individual Muslim, hurts us collectively as Muslims at times when we need to speak with one voice. But if we established an authority figure who speaks for all Muslims, we would be defying a core value of Islam.

I am asked all the time why Muslims haven't condemned 9/11, why there is a "deafening silence" from the Muslim world, failing to speak up against terrorism. Everyone from my neighbor to Mort Kondracke of FOX News has asked me, and many more unknown to me continue to propagate the untruth, designed to smear the innocent Muslim people of the world, implying that Mulsims do not condemn the attacks of 9/11. It's been seven years since 9/11, and I am still constantly asked this question. Mort asked me when I was speaking on a panel in October 2007 at Washington, DC's historic Meridian Center. I expected some questions beyond this typical one, but Mort insisted to me and the two other speakers on the panel—the Georgetown University imam and the press secretary for the Jordanian embassy—that Muslims hadn't condemned the 9/11 attacks, despite the fact that three of us were sitting next to him assuring him that not only had we condemned them, but that every Muslim we knew had. I told him to do a Google search on "Muslims condemn 9/11," and he would find multiple pages of condemnations from Muslim leaders and average Muslims. But he then said he would like to see a full page *New York Times* ad of Muslim leaders condemning 9/11. I just sighed to myself and sat back. Some people will never be convinced. The myth of noncondemnation will forever haunt Muslims and comfort Islamophobes in their persistent bias.

The pre-Islamic pagan tribes of Mecca worshiped 360 gods. Idols for each one sat in the Kaaba, which, although originally built by Abraham in devotion to the one God, had now been commandeered by pre-Islamic Arabia's pagan religion. Reportedly, each tribe had one god it worshiped out of the 360. One's religion and spirituality then was bound to one's tribal background and ethnicity. You could only access the god of your tribe. Furthermore, if your tribe happened to be a weak one, your god may have been seen as a lesser god compared to a more powerful tribe's god. Religion had its own politics. The leader of the tribe probably considered himself to be the most in touch with God, and then his subordinate, and then that person's subordinate and so on.

Islam changed the politics of religion. The pronouncement of Islam gave the Meccans the chance to access God as individuals and not through one's tribe. You had your own right to speak to God. You did not have to be in a tribe to be entitled to worship God. "It is We who have created man, and We know what his innermost self whispers within him: for We are closer to him than his neck-vein." (50:16). You don't have to go through your superior, then your superior's superior, and then wonder about what happened to you. You go straight to the Boss, and the Boss will never fire you. As a Muslim, you and God deal directly with each other: "And strive hard in God's cause with all the striving that is due to Him: it is He who has elected you [to carry His message], and has laid no hardship on you in [anything that pertains to] religion, [and made you follow] the creed of your forefather Abraham" (22:78).

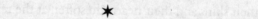

*

Other religions cite miracles as proof of God's presence and existence. Islam is a religion with few miracles compared to others. In Catholic school, I think I must have learned about a new miracle every week. The grade school I attended, Saint John Neumann Catholic School, was named for the first American saint. He lived in Philadelphia, and he particularly loved children. His only known miracle was his curing of a deathly sick child. *What is the miracle of Islam?* I used to wonder. I had never heard of Muhammad healing sick children. I made a mental note (or as much as I could at the age of seven) to instruct my family and friends to pray to Saint John Neumann if I ever became sick before I grew up.

The miracle of Islam, though, is the Qur'an. Scholars and Arabic readers universally agree that the Qur'an has the most beautiful and touching language in Arabic. Many people who read the Qur'an in Arabic, Muslim or not, say that the Qur'an must be divine because its writing is so incredible. In fact, Muslims are required to learn Arabic so that they will read the Qur'an in its original language—how God intended it to be heard. Reading the Qur'an is supposed to be a powerful experience, charged with emotion and excitement. Muslims believe that the Qur'an, as the word of God, is not simply a text but an event unto itself. Even the Qur'an describes itself as an earth-moving experience: "God bestows from on high the best of all teachings in the shape of a divine writ fully consistent within itself, repeating each statement [of the truth] in manifold forms—[a divine writ] whereat shiver the skins of all who of their Sustainer stand in awe: [but] in the end their skins and their hearts do soften at the remembrance of [the grace of] God" (39:23).

The reading of the Qur'an is a divine experience for Muslims, a real-time encounter with God. Many Muslims describe reading the Qur'an this way. They encourage others not to read the Qur'an in a rush but slowly so that every word can be savored. Simply in reading or hearing a recitation of the Qur'an, you can feel God. Reportedly, some of the Arab pagans, upon hearing the Qur'an for the first time, converted on the spot.

A member of Muhammad's tribe, Umar ibn-al Khattab, was a known enemy of Muhammad's new movement. He was a devout pagan, committed to his ways. He allegedly even plotted to kill Muhammad. As legend goes, he was returning home, when he heard a recitation coming from inside his home. He tore open his door to find his sister, who had secretly converted to Islam, listening to a Qur'an reciter she had invited to their home. He was so angry that the Islam he hated had found a way inside his home that he flung himself at the reciter. In the commotion, Umar mistakenly knocked his sister, Fatima, whose back was to him, to the ground. She turned to look at who had knocked her over, and when she looked up at Umar, he saw that she was now bleeding—an injury from the fall. Umar felt ashamed for hurting his sister. He may have been a pagan and hated Muhammad, but he didn't mean to be cruel to his sister. The reciter had also dropped the passage he was reading, which Umar immediately picked up. He was a scholar of Arabic poetry and one of the few literate men of Mecca. When he read the Qur'anic passage, the beauty of the Qur'an's words stunned him. It was like a jolt to his system. Even in all his reading and scholarship, he had never read anything like this. A convert in spirit already, he formally converted later and became one of the heroes of early Islam.

A second story also describes Umar's conversion and is similar overall.* Basically, Umar surreptitiously listened to Muhammad's own recitation of the Qur'an one evening. He hid himself behind the cloak hanging over the Kaaba, where Muhammad was reciting the Qur'an late at night. Without the eyes of Meccan society on him, Umar was free to listen to the Qur'an minus the knee-jerk, anti-Islam rhetoric. As in the first story, he found himself riveted by the beautiful language. Umar began to cry at hearing it. The language of the Qur'an, so beautiful yet so simple and clear, moved him.

---

* For both versions of this story, I rely on Karen Armstrong's *A History of God: The 4,000-Year Quest of Judaism, Christianity and Islam* (New York: Ballantine, 1993), 145–6.

Many converts I've met have the same story. They only began reading the Qur'an because they hated Muslims and Islam. Unaware of its power, the Qur'an crept inside them to a place they did not know existed. They saw a truth—the light and the compass that only people who are "found" religiously can understand.

My grandmother had a long, drawn-out deathbed experience. Many family and friends came to visit and sit with her. There was a copy of the Qur'an in her room, and many would open it and read to themselves in Naunni's presence. A relative of mine, whom I had previously steered clear of due to his strict Wahhabi views on Islam, joined me and my brother as we were sitting vigil for a turn. I would have preferred that this relative not join us. I wanted to spend the time alone with my grandmother and not risk a lecture on whatever fundamentalism I was disregarding in my life. The spirit of the occasion didn't permit me to kick him out though.

As I was resigning myself to the situation, this relative asked if we would like him to recite from the Qur'an. My brother and I looked at each other in bewilderment—such an offer certainly didn't seem like an invitation to a religion lecture or argument. One of us said yes, go ahead if you want. Nervous at what he meant to do, my brother and I were tentative as our relative reached for the Qur'an and opened it. He swallowed, took a deep breath, and began reciting. He had a nuanced but disciplined voice, which originated in his stomach, was styled by his heart, and finally delivered through his throat. The words filled the room's bleakness. I felt like invisible arms were hugging me.

He recited for some time and then eventually took a break. Later on, when he was leaving for the evening, I put my arm around his shoulder to say good-bye and told him how comforting his recitation was. He was modest and said that his teacher was much better than he and that he hadn't practiced in a while, but I just held him a little closer and said that I hoped he would do another recitation the next time he visited. We had also sat together that evening and talked about religion, and I realized he had changed his strict views to more moderate ones. I actually enjoyed talking to him. I can't help but think that the voice of God, speaking through the Qur'an, renewed some love between us. It was the Qur'an and the feeling of calm that its words gave me that brought us together. I had spent years avoiding this relative, and suddenly I wanted his company. The Qur'an causes daily, small miracles like that.

He [illegible] the time [illegible] them [illegible], took a [illegible] on [illegible] of [illegible] and [illegible] around his shoulders [illegible] good-bye, and told him how [illegible] was. [illegible] and said that his [illegible] at the [illegible] hold him [illegible], and said that [illegible] he would [illegible] the next time he [illegible]. We [illegible] and [illegible] [illegible] I actually [illegible] and I can [illegible] the voice of God [illegible] the [illegible] gave me that I could [illegible] years [illegible] daily [illegible]

Chapter 3

Sufism: A Rich Mystical Tradition

Each interprets my notes in harmony with his own feelings,
But not one fathoms the secrets of my heart.

Maulana Jalaludin Muhammad Rumi

When I was nineteen years old and a freshman at Wellesley College, I lived in a dormitory neighboring our campus lake that shimmered even on the coldest New England winter days. The tradition was that if your beau accompanied you on a walk around the lake three times, he must then ask you to marry him. If he didn't, you were free, even under Wellesley etiquette, to push him into the lake. In nearby Green Hall, I would attend my first college class (of many) on Islam. The woman who would later become my advisor was lecturing us on Islamic mysticism.

The Sufis, she told us, were Muslim mystics—poets, musicians, artists, writers, and others—who had a deep love for God. They first emerged in the mid-eighth century in response to the then Muslim leadership's focus on political affairs. The Sufis wanted to remember Muhammad's emphasis on unity and brotherhood with members of other religions. The Sufis also longed to be with God, and they felt that no one really understood them the way God did. They had many ways of becoming closer to God. The artistically inclined ones would work themselves up into ecstatic raptures—through dance or poetry.

The famous thirteenth-century Sufi poet Rumi once compared himself to a reed growing by a lake that was ripped from the ground to be used as a reed flute. Beautiful sounds came out of the reed, which resonated with each listener, capturing each one's personal feelings. But the reed wanted to be back in the earth, near the lake. "Like the reed," my teacher told us, "Rumi wanted to be returned to God. His love for God was that strong."

Hearing that poem for the first time, I both understood it and didn't. When I left class that day, I felt as if I were seeing everything for the first time. I saw the fingerprints of God everywhere—in every tree leaf, in the blades of grass, and even in the glow of the lake. Some Sufis say that they are actually in love with God and that only God knows them intimately. Once you realize, as Sufis do, that God is in everything, you can't help but always look for Him.

Sufism is the mystical attitude or phenomenon within Islam. It is probably the aspect of Islam that is most attractive to non-Muslims because, generally, Sufism espouses a philosophy of listening to one's heart and emotions to feel the presence of God, rather than, for instance, focusing on strict rules. While all mystical traditions are experiencing a renaissance these days, like the Kabbalah movement of Judaism and the tradition of Catholic mysticism, Sufism has been a catalyst in the Islamic world since it arrived on the scene. Its message is clear and spare, minimalist almost: focus on God.

Every time I am confronted by Sufism's clarity—whether through a Rumi poem or a Sufi folk song—I am stunned by it and seduced by its simplicity and single-minded focus on the one God. Sufism is not a specific sect or branch of Islam but actually cuts through all the various schools and sects. As a result, one can be a Sunni Muslim and also Sufi or an African-American Muslim who holds Sufi ideals, and so on. One of my reviewers for this book, Dr. Maher Hathout, who is a scholar of the Qur'an, remarked after reading this chapter, "As a matter of fact, there is a Sufi inside every believing Muslim." Furthermore, while the essence of Sufism is the same from Sufi to Sufi, individual Sufi orders maintain different practices and traditions for worship, passed down for hundreds of years.

The converts to Islam I meet these days either converted because of the lyrics of avowed Muslim rappers like Mos Def or Q-Tip or because of Sufism. They'll tell me how they found a book of poetry by Rumi or Hafiz (another famous Sufi poet, who lived about one hundred years after Rumi) and became immediately hooked. In fact, since the beginning of the Sufi movement, Sufism has assisted the spread of Islam.

Contrary to widespread popular belief, Islam was not really spread by the sword—most Muslim leaders including Muhammad abhorred forced conversion. The Qur'an itself says, "Let there be no compulsion in religion" (2:256), and in my experience, Muslims do not engage in missionary activity. (Judaism has a similar prohibition on missionary activities.) However, one could say that Islam was spread by Sufism—not in all areas of the Islamic world but certainly in South and Central Asia. Muslim conquerors might have gained power in a certain area, but the appeal of the emotion-guided Sufis actually caused the locals to convert or to become more devout Muslims. Even near-illiterate Pakistani rickshaw drivers can sing a Rumi poem to you while they transport you in their colorful conveyance. In South Asia, it is not even uncommon to find a Hindu or Christian praying at the shrine of a Sufi saint; the appeal of Sufism is so strong that it stretches beyond formal religious boundaries.

It's easy to talk in generalizations and allusions to Sufi philosophy, but I really think it works. I know it does because I have actually tested Sufi philosophy on myself. In promoting my first book, I reached out to many people, including people in the media. The more forward movement I had, the more publicity I would receive. When one story idea fell through, another would come in, as long as I kept trying. I was once being courted by a major television network. I had stars in my eyes and dreams of greater things that would make my family and friends proud. All the signals were there. It looked like it was going to be huge news.

But then it didn't work out. It was a quick finish to a long process, where I had jumped through hoop after hoop. I told myself that I had to accept my life as it was—that I was a one-book author, and that was great. I kept doing op-ed writing and my regular attorney job and would let the future play itself out as it would.

Previous to this rejection, an agent who had opted not to represent me, despite a strong reference, contacted me again, about a month after the television network item ended. Would I be interested in writing a book called *Why I Am a Muslim,* he wanted to know. A publisher had asked him if he knew of an American Muslim woman who could write. I said yes, and soon I was writing my second book. As Sufism instructs, I had kept moving forward and not let the setbacks—including when this agent previously rejected my proposal—devastate me. I was still around then when the positive energy I had put out into the world was ready to come back to me. I hadn't become so dejected that I turned myself off from the world because of a couple of rejections. I stayed in and stayed open.

What makes Sufism so appealing? Several things. First, Sufis are very inclusive and open-minded. They believe that God is everywhere. They quote Qur'anic passages like, "Wherever you turn, there is the face of God" (2:115). If God is everywhere, then God can be seen in all religions too. A Sufi would not only be comfortable praying at a church or synagogue or even a Hindu temple but would also be comfortable ministering to a non-Muslim. Furthermore, because God can manifest in different ways, according to the Qur'an, Sufis are particularly open to new experiences or feelings: "Every day He manifests Himself in yet another wondrous way" (55:29). Every day, God is different, and the Sufi is on a constant quest to find Him.

Sufis love everyone and everything, regardless of who or what they are. Religion, wealth, skin color—nothing matters.

Sufis feel that Muhammad lived without any prejudice and was able to achieve religious enlightenment because he was so open. So they themselves want to be open too.

A Sufi feels that if he gives love, then he will be able to accept easily the love that exists created by God. Everyone can be redeemed because good/God exists in everything. God is bigger than any one religion, and although Sufis feel that Islam gives them the proper framework to access God, a different framework may work for someone else. As Hafiz writes:

I
Have
Learned
So much from God
That I can no longer
Call
Myself
A Christian, A Hindu, A Muslim,
A Buddhist, A Jew.

The goal of the Sufi, generally, is to become fully absorbed into God. All the unimportant matters—the religious divisions, the politics, the "I look fat in this dress"-type concerns—fall away. A Sufi, too, needs to keep himself or herself relaxed and unstressed. One can't grow internally otherwise. A good boxer knows that to throw the best punches, he has to stay relaxed and not tense or his body will betray him. A Sufi feels the same, but about God. Focusing on internal growth keeps one open to religious enlightenment.

What is left is a shell that, because God created it, will be able to reach God easier than with all the trappings of human life layered onto it.

Sufis pay attention to current events and politics, but they keep a distance from them, too, because they know that God cannot be found by focusing on the news. The Qur'an tells us that God placed a small bit of divinity inside each of us when we were created. Exploring ourselves and trying to reach that bit of divinity will be more fruitful in our quest for spirituality than in reading the *New York Times* cover to cover or watching CNN.

The Sufi has a long view of history and the world. Junoon—Asia's biggest rock band—is heavily influenced by Sufism, as many Muslim artists are. They coined the phrase "Sufi Rock." One of the classic Sufi songs they have redone (and also sing in Urdu, the language of Pakistan) emphasizes in its lyrics how the Sufi sees the world:

When this earth was not here,
There was no sky.
There were no moon and stars.
The only thing here was
You, only you, only.
Allah Hu Allah Hu Allah Hu
Allah Hu Allah Hu Allah Hu.

Similarly, a Sufi must be patient. The rewards of Sufism will not be precipitated by a furious chase after them. God can be found by focusing on oneself because, just like all of God's creations, God is inside you—a theme Rumi wrote much poetry about:

> *You have woken up late,*
> *lost and perplexed*
> *but don't rush to your books*
> *looking for knowledge.*
> *Pick up the flute instead and*
> *let your heart play.*

When I was little, my mom and usually my younger brother would pick up my sister and me from school. Most days we would head to the bank and then the grocery store—our neighborhood Safeway. Pakistani cooking requires a lot of fresh ingredients. They are key. A growing family of five eats a lot of food too. My brother sat in the grocery cart in the seat with the leg holes, which was a good thing since he tended to run off when not restricted like this. I hung off the side of the cart, with my feet planted on the lower tray—my instep surfing the rounded edge and my forearms resting on the rim of the large basket. My sister would wander back and forth from our cart to other parts of the aisle or would be sent by my mom to find some special item. We were a little galaxy with our mom as our sun, and us like planets orbiting around her. We bought flowers, too, about once a week, but I didn't have much interest in this part of our trip. It seemed pointless to buy flowers; they'd just die in a few days anyway.

My mother usually carried the flowers into the house herself while my sister and I wrestled with the rest of the groceries. She would first wash the produce items and put several pots of something on the stove to cook. Then she'd turn her attention to the flowers. First, she would fill a vase with water and place it near the sink. She would then put all the flowers in the sink. She would take off some of their leaves. Then she'd remove the outer petals that had brown spots. She'd cut the thorns off the roses. Finally, with her kitchen scissors, she would cut off the stems' ends under running water. One by one, she would trim each flower and then place it to the side of the sink. All the scraps went into the trash.

Next came the most perplexing part. She would put the trimmed flowers into the vase one by one, studying the arrangement as it developed. When it was done, she would stand in front of it like she was contemplating—sort of the way I would later imagine da Vinci standing in front of his portrait of the *Mona Lisa*. Often, she would take out all the flowers and start over. She would rotate the vase and walk around it, looking at it from all angles.

I did not understand what she was doing or why for years.

Sufism is generally not political. A true Sufi wants to develop his or her own soul. She will even isolate herself from other people and fast for days, hoping to arrive in a divine frenzy where she feels the presence of God, the creator.

In the same religion class where I learned about Rumi, my professor taught us a little about the whirling dervishes. She explained that they are Sufis, based out of Turkey, who feel that when they spin (or whirl) for long periods of time, they reach a level of consciousness—a "high" of sorts—where they can feel God's presence. Students form rows around the *shaykh,* who is their teacher and spiritual leader.* The *shaykh* represents the sun, and his dervish-students, the planets of the solar system orbiting around him. Like my own family at Safeway, the dervishes are their own universe, with its own harmony and sense of spirit. They can spin for hours, even days.

Sufis also feel that experience and encounters are of more value than *reading about* experiences. Whatever activity is going to produce the most awareness of God is preferred to activities that won't.

Many non-Sufis seem to know about the whirling dervishes. In fact the first time I saw some was in a Madonna video. They were men in exotic outfits with tall fezzes and long purple robes. They were spinning repeatedly in small circles—the whole group of them, over two or three dozen at least. I instinctively knew they were Islamic.

It all sounds very exotic and odd, doesn't it? Especially if Madonna is interested in it. But Muslims grow up exposed to everyday acts of Sufism and realize its place. Sufism can be a daily part of one's life like anything else, or it can be the directional compass by which one lives one's life. Today, many

---

* A *shaykh* is a Sufi Muslim spiritual or religious leader. In order to earn the title, one must study under another *shaykh* for a certain period of time. In some families, the title is also hereditary. *Shaykh* is also spelled "*sheikh*" depending on the *shaykh's/sheikh's* preference.

people engage in recreational Sufism, but a Sufi can easily be as hardcore and as fervent as a conservative Muslim.

When my mother was a little girl growing up in Pakistan, she visited the home of her paternal great-aunt and uncle. On one end of the house was a room with a door that was almost always closed. One time, my mom saw the man who lived in the room. My mom didn't know why he had his door open, but he had a long, graying beard and a long blue tunic. The viewing was so brief and so rare that my mom is still not sure if she saw him or just made up the memory based on what she has heard about him. It was her uncle—the brother of her paternal aunt, and he was a Sufi.

He stayed in his room for most of his adult life, meditating and praying and trying to reach God. He was called "Sufi Jee," which literally means Mr. Sufi in Urdu, the language of Pakistan. Magically, trays of food that had been left for him would reappear outside his closed door—the food eaten and the dish and silverware gleaming with cleanliness as if food had never been put on them. He was famous all over Pakistan, and people came from far and wide to see him, my mother says. They thought that if he could stay in his room all day, he must be blessed by God. To visit his home was almost holy. People would meet him and ask him to pray for them. When prayers were answered, the people returned, although Sufi Jee could not be claimed as the source necessarily. He never left the room till a few years ago, just before his death. Living most of his life in this room did not bother him, because a Sufi is not concerned with the daily events of life.

A few years after graduating from Wellesley, I found myself back in New England—this time at Harvard. I had been asked to appear on a panel at a conference on Islam in America, and there, sitting next to me, on the end of the panel, was the Sufi Shaykh Taner Ansari, who is one of the authors of a book called *The Sun Will Rise in the West: The Holy Trail* and has a Sufi school in Napa, California. I knew he was Sufi because he was dressed funny. (At Muslim gatherings, you can usually spot the Sufi *shaykh*. He's the one who looks like he's dressed in costume.)

Shaykh Taner had on a pair of pants and a waist-length black robe that he had wrapped around himself and tied with a waistband of the same fabric. He also had a matching prayer cap on his head. His robe and cap were made of wool—the same wool the Sufis were named after.

In the ninth century, a group of Muslims, like this *shaykh,* began wearing the course woolen garments Muhammad is known to have worn. They wanted to emulate the Prophet Muhammad, not only because he is a model Muslim and person, but also because the community felt they had lost touch with his example, focusing too much on politics and the strict practice of Islam. Muhammad was religious, but they felt he did not focus on rules and regulations. To them, the still-young Muslim community was focusing on these rules to the detriment of the inner spiritual development Muhammad had perfected. Muhammad wore *tasawwuf,* which was a kind of wool, because the poor people of the time wore the same wool. Muhammad wanted to show that all Muslims and all people—rich or poor—lived and worshiped as equals. To symbolize their connection with what they felt were the original goals and mission of the Prophet, this group wore the *tasawwuf* fabric, and from *tasawwuf* the word *sufi* was derived.

Shaykh Taner had fair skin and light blue eyes. He was an older man and was so calm that at first I was uncomfortable. When the time came for his speech, Shaykh Taner was suddenly invigorated. The mild-mannered Sufi had become the hyperactive speaker. It was a good thing he spoke behind a podium because otherwise all his energy might have hurtled him forward and into the audience. My speech came next. I was introduced as the Muslim Feminist Cowgirl. I spoke about the reaction to my first book and how the American Muslim community needed to be more unified and less critical of one another.

After we had all spoken, the entire panel moved outside where those in the audience could meet us and ask questions.

"How dare you say that you are a Muslim feminist?" a young American Muslim woman in a turbanlike *hijab* and skintight jeans was yelling at me. I was quite surprised to be the object of her scorn. She felt that the phrase "Muslim feminist" was redundant since Islam already incorporates the principles of feminism. To call myself a Muslim feminist would give non-Muslims the impression that feminism did not exist in Islam. She did have a point, I thought. So I said to her that I only use the title so that non-Muslims will know that one can be a feminist and a Muslim at the same time. I did not mean to take anything away from Islam. She was not satisfied, though, and I wondered if she expected me to yell back at her.

"Listen, she has heard what you are saying. You are welcome to write your own book. Now let that be the end of it," Shaykh Taner's slightly accented voice piped up. He had been sitting next to me, listening to this woman for as long as I had, but I had no idea he cared. I realized the *shaykh* was onto something. To her endless barrage, I just kept saying, "You heard him. I encourage you to write your own book." Finally, my fan gave up and left. I didn't need to yell back so much as defuse.

At that very moment, the conference organizer began yelling out that the last shuttle vans from the Science Center would be leaving soon. It was evening and a cold March in Cambridge, and I did not want to miss the shuttle off the massive Harvard campus.

"Don't let it go without me," I yelled back at the organizer, who was already busy with something else.

"Don't worry," the *shaykh* said to me gently. "You will have a ride home." I wanted to say, "Thanks for the rescue back there, but I don't think I can count on God to find me a ride to my hotel, Shaykh." But something about his demeanor told me not to say anything. I still had people waiting to talk to me about my speech, and the *shaykh*'s student and his wife were beginning to gather up his things.

I had left myself in the hands of a man whom I didn't know at all. He would periodically look back at me, hold his palm up at me the way a crossing guard would, and say, "Don't worry. We are not going anywhere." When I finally finished speaking with everyone, I realized that the *shaykh* and his group, which included another speaker, were waiting for me. We all headed off to the small car of a student of the *shaykh*'s and sped off to the hotel. The *shaykh* had kept his promise to me and didn't once complain that I had delayed his departure by nearly forty minutes.

Shaykh Taner writes in his book that a true Sufi should be of such a nature that when you are around him, you forget your troubles—that the Sufi takes on your problems for you. Furthermore, because the *shaykh* himself is at peace, he will not appear to be troubled by your troubles, but you will have the benefit of being freed from them.

I had Guatemalan "worry dolls" when I was little, which served a similar function, albeit on a much smaller scale. They came in a little plywood container not much bigger than a quarter. They were tiny stick figures, probably made of a papier-mâché–type substance and dressed in tiny, bright clothes made of hot pink, white, and yellow string. As the legend goes, you would tell each doll one of your problems just before you went to sleep and leave it out of the container. As you slept, the doll would work on your problem for you. Guatemalan spiritualists and Sufis probably both know that, in time, most problems resolve themselves, especially if you can unburden yourself of them enough to focus on how to solve the problem rather than worry about it.

Shaykh Taner—whom I had never met before and who had no reason to assist me—had taken my problems away from me. The Sufi way to deal with the young woman with the strident comments was not to yell or engage in a shouting match, but simply to say thank you and acknowledge her comments.

The Sufi way to help another is not to fuss or draw attention to your assistance but simply to help. In one evening, Shaykh Taner had helped me twice, and without any heavy lifting. When the time came for him to be hyper, in delivering his speech, he was able to tap into the reserve of energy a Sufi has. A Sufi is guided by his emotions, but he never lets his emotions control him. He controls them for maximum effect when needed.

A real Sufi is always content in all circumstances: rich or poor, hot or cold, sick or healthy. To become a Sufi, you must practice self-control to relax yourself through music, concentration, or meditation. Once you can control yourself so that you do not feel the upheavals of emotions, you can channel your positive energy and maintain an even balance. As a result, the Sufi is always happy and can call upon his or her stored energy when needed. Shaykh Taner went from a constant, small flame to a ball of fire during his speech and then back again afterward.

The Sufi changes his or her mind-set to fit the results with the sense that God must have meant for these results to happen. The Qur'an says: "But it is possible that ye dislike a thing which is good for you, and that ye love a thing which is bad for you. But God knoweth, and ye know not" (2:216).

My mother's family has strong Sufi leanings, not just because of her uncle Sufiji but also because of other Sufi members of her family and their frequent use of Sufi reasoning and attitude.

My mother fully became a Sufi the day she returned to Pakistan after a year of marriage and another year of married life in the United States She was twenty years old. Immediately upon arriving in the city of her birth, Lahore, she went straight to her trunk. In the trunk, she had saved everything that was important to her from her unmarried life—her favorite books, her awards, her magazines, her journals, even a scrapbook with pictures of her Pakistani movie heartthrob. She couldn't wait to see it. When she opened the trunk that day, she was stunned to find the trunk completely empty. My mother went into shock. She couldn't breathe. She felt as if all her memories were gone. She finally ran to her mother—my grandmother, hoping that perhaps she had put away her things for safekeeping. "Well," her mother said, "Your brother and sister took what they liked, and I sold the rest to the recycling man." The recycling man came around to the neighborhood homes about once a month to buy any valuable trash. He would also sell any items of his that you were interested in. My mother was so angry. Who did it? She interrogated my grandmother. "Your brother and sister, I told you!" she snapped back. In response to the same question, my mom's siblings said "Mom did it. We didn't touch it." After hours of crying and turning her parents' home upside down, she found three or four items only. The recycling man had the majority of the booty. Of course it would be impossible to find the items now. Once the recycling man took them, within hours they could be anywhere on the Indian subcontinent. She knew that Sufis believed in focusing on God and not being distracted by materiality. A determined young woman, she resolved never to be so controlled by her possessions again.

She never again wanted to be so attached to anything material that losing it would cause her such anguish. From that day on, she held my older sister, who was just one year old then, a little tighter, hugging her and kissing her more often than usual to make up for her loss. She had become a Sufi. She doesn't collect things anymore, but she treasures what really matters.

Perhaps because of the Sufi influence my mother was exposed to, we always reasoned our way out of anything that made us unhappy. If you did not receive something you really wanted, then it was good that you did not receive it. Maybe if you had received it, something bad would have happened to you. Or maybe you would not be as happy as you are now.

Because of Sufi philosophy and my own life experiences, I've learned that the biggest disappointments are almost always followed by even greater successes or happiness. At the very moment of the disappointment, of course, I am as annoyed or mad or sad as the next person, but soon after, I find myself giddy with excitement at the prospect of what is coming next. Sometimes, I almost look forward to disappointing news not only because something bigger is just on the horizon but also just to see how my family is going to explain the disappointment Sufi-ly. This one aspect of Sufism is so powerful that, by embracing it, you may never have any real regrets in your life.

★

I used to write a consciousness-raising column for a Pakistani American newspaper. I chose to devote one column to the story of a young, gay Muslim man's struggle to be both homosexual and Muslim. I wrote the piece carefully, having interviewed the young, gay Muslim for hours. I opened with a description of how he had been stopped by the FBI while coming off of a plane soon after the tragic 9/11 attacks.

I wanted the Muslim reader to come away with the feeling that we are all targeted for whatever bias or prejudice the targeting person has—whether that's being Muslim, having brown skin, or being gay. If we wanted Americans to respect us as Muslims, we had to respect differences ourselves. I wasn't advocating homosexuality or arguing that it was permissible in Islam but was simply showing that hate of anything comes from the same place. If it's okay for a Muslim to hate a homosexual, then it's okay for a non-Muslim to hate a Muslim. I was trying to show that this philosophy would only hurt our community.

To my shock, my editor refused to print the column! I had never been censored before, and I even briefly enjoyed my renegade status. Soon, I realized that I really *needed* to publish this piece. On a whim, I sent the column to an editor at Beliefnet.com, fully expecting her to send me a nice "Thanks but no thanks" e-mail. Within a day, I had an e-mail from her saying the piece was right up Beliefnet's alley and that they would pay me for it! No one was more surprised than I—being censored from a paper that paid me nothing actually resulted in publication with more exposure *and* being paid! Something that I dislike—censorship—became something that was good for me after all. To dwell on the disappointment or bad news is a failure to see the opportunity. It keeps one closed to greater success and, for the Sufi, to potential religious enlightenment. A Sufi is always open to what will come next.

For similar reasons, Sufis don't care about material things or possessions. They can live with or without them. Whenever a Sufi loses a valued possession or accidentally breaks something, he or she will say, "Lucky ones—it happens to their possessions. Unlucky ones—it happens to their life." It doesn't matter if the antique china plate breaks as long as we are fine. If you are able to master the Sufi philosophy, you will stay happy.

The great singer Nusrat Fateh Ali Khan is probably the best-known modern-day Sufi. He was a *qawwali* singer. *Qawwali* is a Pakistani tradition of singing with a band of musicians who play drums made of wood and goatskin. Ali's songs—many of them old, Sufi folk songs—capture the Sufi philosophy. "Mast Kalander" is his signature song, which translates roughly as "carefree, free, roaming person." The song has a rhythmic beat, and Ali's singing has a dramatic tension filled with promise. It is nearly impossible for me to hear this song and not stop what I am doing and listen to it. I am caught up in the beat and Ali's singing, wondering what will happen next in the sounds of the song. I've heard the song thousands of times, but each time I hear it, I feel like I am hearing it for the first time. The song captures my ears and doesn't let go. A musically inclined Sufi would probably start dancing to the music, taken in to a rapture by it or sway with the rhythm. They feel that God is guiding them to this rapture. Obviously, someone who can be so taken in by music is not worried about whether or not he's lost his leather jacket or favorite pair of shoes. The Sufi places attention on where it's needed, not on concerns that would be frivolous in his or her eyes. In fact, Nusrat himself may have died of his own "carefree" attitude. He had a liver and kidney disease that he never vigilantly medicated or monitored, and eventually it caused his death.

But chances are Nusrat would not have been saddened by his own death. Sufis look forward to death because their mortal existence is ended, and they are finally reunited with God. In fact, their devotees celebrate the deaths of great Sufis, who are called Sufi saints, because the Sufi, in death, has reached his or her goal of reunification with God. Called the *urs* celebration, it draws locals and travelers from far and wide to the shrine where the Sufi is buried. Usually the group, which consists of those whose previous prayers at this Sufi's shrine were answered, prays together all day and night, praying both to God and the Sufi saint. For some Sufis' *urs,* the gathered rock out—dancing and chanting, celebrating the death of their saint and his return to the one God. Similar events are held on the Sufi saint's birthday as well.

The faithful use these *urs* to pray for their worries and concerns and to ask the Sufi saint, who has now, they believe, been absorbed into God, for his or her help. They pray to them to intervene with God on their behalf. Technically, they are not praying to the saint but asking the saint to be their attorney before God and convince God to grant their prayers.

In the late fifties, my grandmother prayed for a son at the *urs* celebration of the Sufi patron saint of her family, Baba Farid Shakur Ganj. "I've had two daughters already, and now I want a son," she prayed. She stayed up all night praying to Baba Farid, who is also called the "Shukur [Sugar] Sufi" because of his odd consumption of brown sugar. Legend has it that he wanted to spend as much time praying and meditating as he could. To minimize visits to the bathroom and the need to bathe (theorizing that smelly food results in the need to bathe), he would eat only brown sugar for days. The sugar provided enough nourishment to keep him alive but also keep his focus and time on God. He is said to have hoarded sugar for this purpose.

Once, a group of local traders with a caravan full of sugar came across his path. When he asked them what they were carrying, they of course did not say sugar, as they knew he would want it all from them and they would lose their investment. "Salt," they told him, and he let them go by. When the traders arrived at their destination and began selling their caravan's load, they discovered that they did indeed now have a wagon full of salt and *not* sugar. They ran back to the Sugar Sufi and asked him what he had done. He explained calmly that they had told him they were carrying salt—so that is what they were carrying. My grandmother had two sons over the years after her all-night vigil on the anniversary of the Sugar Sufi's death. (At last report, they both have a taste for sugar as well.)

✶

Sufi shrines play a large role in the births of Muslims and non-Muslims in South Asia. Many South Asians, including non-Muslims, pray at these shrines for a child. My own grandfather claims his birth was the result of his father's praying at the shrine of Salim Chishti in Fatephur Sikri in India. The grave of Chishti, who was the Mughal emperor Akbar's advisor, is enclosed by a carved marble latticework square. You can see through the latticework, and light peeks through the intricately carved openings. Pilgrims come from all over the world, making wishes and requests. Locals pass out small pieces of string to the pilgrims, who then make their wishes and tie the strings to openings in the lattice wall.

When my grandfather was born, his father could not make the journey back to Fatephur Sikri. But he asked another pilgrim to remove a string on his behalf. Whenever a request is granted, you must remove one string. It doesn't have to be your original string. In fact, it would probably be impossible to remove your original string. When I visited the shrine in 1997, I saw thousands of strings knotted to the walls. The loose ends flapped with the slight breeze, and hundreds of pilgrims milled in and out of the shrine. The only place in India I saw that had more visitors was the Taj Mahal.

Naturally, a lot of non-Sufi Muslims, upon hearing these fanciful philosophies and practices, say that Sufis are quacks. The Sufi practice of worshiping or praying to saints that even non-Muslims pray to is particularly troublesome. In Islam, God has no intermediaries, and to worship anything besides God is *shirk* (associating partners with God). Worshiping anything that is not God is probably the only act that all Muslims—conservative and liberal—would agree qualifies as *shirk*. However, to the Sufi, another Sufi who had achieved a high level of peace must have, in order to have achieved that peace, been in closer union with God than the average person. So by praying to that particular Sufi or praying to God at that Sufi's shrine, you would have a better shot at gaining God's attention. In addition, those who "pray" to Sufis feel that they are not praying but simply asking the Sufi saint to intervene with God on their behalf.

At the same Harvard conference where I met Shaykh Taner, I witnessed blatant attacks on Sufism by other non-Sufi Muslim academics. Sometimes Sufis were on the same panel as the person attacking. To these non-Sufis, the Qur'an is a book of guidelines, rules, and parables, not just a springboard into what they view as a spiritual grab bag. I think the critics would do well to look past the obvious *shirk* problem to the value that Sufism has added to Islam. For instance, Shaykh Taner's book quotes the Prophet Muhammad as saying in a *hadith* that spending an hour in academic discussion is better than praying all day. The idea behind this *hadith,* and in some of Sufism, is that enlightenment and greater awareness of one's religion is important. Prayer without understanding the prayer is useless. Rumi, too, criticized a blind focus on rituals and rules without understanding their meaning. In one of his poems, he addressed pilgrims returning from the *hajj* to Mecca. He asks them:

You rave about the holy place
and say you've visited God's garden
but where is your bunch of flowers?
... There is some merit
in the suffering you have endured
but what a pity you have not discovered
the Mecca that's inside.

In addition, Sufism has been responsible for introducing Islam to more people than has orthodox Islam. In the end, these Muslims may not follow Sufism, but it opened the doors of Islam to them.

Fortunately, because Sufis are Sufis, the criticisms don't really bother them too much. Sufis' specialty, since their beginning, has been to defuse politics because of their deep religious belief. The Sufi shaykh, because of his relaxed attitude, is often contrasted with and compared against the truculent mullah, who advocates a view of Islam based on rules and, more recently, political views.

A mullah is a self-appointed Islamic religious leader. Although mullahs have sometimes studied the Qur'an and other Islamic texts, they can just as often be influenced by culture or politics and have no particular background in Islamic studies. An Islamic community can have one or several mullahs to whom members of the community will go for religious guidance and also to observe daily prayers with. In recent years, mullahs in Islamic countries, especially Pakistan, have taken on a political activist role while still assuming religious leadership. Harvard University School of Law professor and Islam expert Noah Feldman argues that this phenomenon is partially due to the fact that many Islamic governments have banned certain political parties or any opposition, which forced critics into the mosques as the mosques were and still are generally unregulated by the government. Although these mullahs are basically politicians, the "mullah" title causes confusion, giving non-Muslims the impression that the mullah is interpreting the Qur'anic view of a political situation and that all religious Muslims must agree with him. Really, these mullahs are like local city council persons.

Over the years of Sufism, Sufis began telling fables about "Mullah Nasiruddin"—probably a fictional mullah. These fables are to show that trusting oneself over the dictates of the mullah is important. Mullah Nasiruddin is often a bumbling character, and I've been collecting his stories because they are so funny. My favorite is the one about the time the Mullah borrowed a donkey from a neighbor. The Mullah needed the donkey to transport some items. Weeks later, the neighbor inquired about his donkey, as the Mullah had not returned it. The Mullah told the neighbor that the donkey was not with him anymore and that the donkey had left of his own accord. The neighbor, hearing the donkey braying and neighing in the Mullah's shed, insisted that the Mullah was wrong and that he, in fact, had the donkey. The neighbor finally said to the Mullah, "I can hear the donkey myself in your shed so you must still have him." The Mullah replied, "Who are you going to believe—a Mullah or a donkey?" Each Mullah story has a moral, and the moral of this one is to believe yourself. When God speaks to you, you will hear Him as clearly as you can hear a donkey, and do not let any Mullah talk you out of believing what you heard.

Through these Mullah Nasiruddin stories and other aspects of Sufi philosophy, Sufism has been able to provide a check on Islamic extremism. When any person has lost touch with his or her inner soul and voice, the natural, gut reaction is to focus heavily on the outward aspects like politics or rules. A consistent failure to connect with the inner meaning of one's religion or belief system—the big picture—can easily turn into extremism—not just with Muslims but with any philosophy. The Sufi has almost always been present in Islam to serve as the reminder to extremists that Islam is not simply about politics or what is done to Muslims. Hafiz criticizes the mullahlike inclination to worship out of fear rather than out of love in one of his poems:

> *Dear ones,*
> *Beware of the tiny gods frightened men*
> *Create*
> *To bring an anesthetic relief*
> *To their sad*
> *Days.*

The point of spiritual development is not to say, "What's the point? I am going to die anyway!" Out of love for the one God who created her, the Sufi welcomes the reunion with God that death provides. Out of love for God, Sufis keep themselves open to new experiences and encounters, which all hold the potential for new enlightenment. The Sufi, by stripping away the layers, will make herself open to God:

Lovers trust in the wealth of their hearts
while the all-knowing mind sees only thorns ahead.
To wander in the fields of flowers
pull the thorns from your heart.
Maulana Jalaludin Muhammad Rumi

When I lived in the Bay Area, I liked shopping at the Potrero Hill Safeway in San Francisco, although I didn't live near it, because it had a parking lot and because it's very similar to the one in my hometown. I was a lone planet with just a grocery cart and no moons. At the end of the store, by the produce, is the flower section. I would buy a bouquet of yellow roses, just starting to open, their petals protectively embracing their centers but still showing the promise of splendor. At home, after I had dragged the groceries in myself with no sister to help me, I would attend to the flowers.

To be a Sufi, you must be open to God and to receiving enlightenment. You must trim off your petals that are already dying and cut off the stem ends to provide a fresh opening for water and nourishment. Then cut off the thorns, from where bacteria can enter the flower and kill it before it blooms. The leaves should be taken off if they do not add to the arrangement, and you should have left before you, as I do, a bouquet of perfect flowers, needing only water and some flower food to bloom. I look at the vase and rotate it. It's perfect the way it is. No need to rearrange. They may die in seven or eight days, but at least they will have bloomed.

Chapter 4

We Are All Imperfect

He Who has created seven heavens in full harmony with one another:
No fault wilt thou see in the creation of the Most Gracious.
And turn thy vision [upon it] once more: canst thou see any flaw?

Qur'an 67:3

"We have all returned from lesser jihad to the greater jihad." Some companions asked:
"What is the greater jihad?" He replied, "Jihad against the desires."

Hadith (saying) attributed to Prophet Muhammad*

"Look down there," our tour guide said to us when we were inside, pointing toward Mumtaz Mahal's tomb. We were inside one of the world's greatest structures: the Taj

* Although the reliability of this *hadith* has been questioned, the importance of the personal jihad over the external one was a common understanding among Muhammad's companions.

Mahal. "See the black spot?" our tour guide asked us. He was a petite Indian man who wore an ascot and navy blue blazer, just like an Indian Thurston Howell III from *Gilligan's Island*. My family and I were visiting India. It was December 1996. "Thurston Howell" told us about how the Mughal emperor Shah Jehan called artisans from all over the world to build the beautiful white marble monument to his beloved wife Mumtaz. Down very low was a perfectly round black marble circle, standing out from the gleaming white marble floor.

"Shah Jehan deliberately had this black circle put there because he did not want to compete with God," he told us authoritatively. He went on to explain that Shah Jehan wanted his wife's sanctuary, his monument of love to her, to be as perfect as possible—totally perfect and symmetrical down to the last nail. A piece of marble with even the smallest imperfection was discarded—and also broken down into pieces so that it could not mistakenly be used. Shah Jehan probably designed the Taj Mahal himself, as no records of any architect exist. He built the palace on a segment of land that was higher than the surrounding land and bordered on a river so that when you looked at the Taj Mahal, you would see nothing but the palace and blue skies framing it—a grand sight. But Shah Jehan, a faithful Muslim, did not want to offend or challenge God by making a perfect creation. So he purposely added a black spot. It had no coordinating black spot. It was a flaw purposely added—in some ways, the perfect mistake.

My father was born in Delhi but hadn't been back since being forced to leave at the age of eight, when Pakistan was created in 1947. He remembered his family home fondly. On our trip to India, he visited his old neighborhood, and though he didn't say it, I think he was depressed at how everything had fallen apart. At one time, his home had been majestic. Because of its large size and an imminent need, his own father had donated it as the schoolhouse for a girl's school. My father's family had also been royally decreed to care for the Rashanara Gardens in Delhi—a former Mughal royal garden, one of which was near his home and held the tomb of a princess. Now things were very different. His old home—the royal garden he used to play in—had fallen victim to Third World deprivation.

On practically the other end of the aesthetic spectrum was the Taj Mahal, which we visited a few days later. We took an early-morning slow train from Delhi to Agra, where the Taj Mahal is located. The train did not fit the tracks well, so when you were riding on the train your body actually vibrated back and forth as the train did. The conductors, who must have been used to this sputtering, served breakfast smoothly, but I could hardly eat with all the vibrations. I would aim for my mouth with the utensil but end up poking myself in the cheek or chin instead. The few morsels of food that did make it to my stomach bounced around in all the rattle. When we finally saw the black spot at the Taj, I looked at it for several minutes. I can't really envision the spot anymore, but I see my own black spots several times a day.

The meaning behind the Taj Mahal's black spot is the Islamic principle that God is perfect. By implication, anything that is not God is not perfect. It is a fact of life in Islam. Shah Jehan, knowing this, did not want to compete with God and blatantly conceded with the black spot. The Qur'an describes God as the ultimate architect: "It is He Who hath created for you all things that are on earth; Moreover His design comprehended the heavens, for He gave order and perfection to the seven firmaments; and of all things He hath perfect knowledge" (2:29).

Humans, on the other hand, are flawed. The Qur'an describes the soul as made by God with the capacity to do right and wrong and also the wisdom to recognize both: "By the Soul, and the proportion and order given to it, and its enlightenment as to its wrong and its right. Truly he succeeds that purifies it, And he fails that corrupts it!" (91:7–10). Because we were programmed to do some wrong, we can take comfort when we make a mistake to know that we are simply manifesting how God made us. I would even argue that by recognizing we have made a mistake, we are already starting to manifest that part of our souls that God made for "right." If we can keep from doing that mistake again, we are also still engaging in "right." In making a mistake, too, we also give God a chance and reason to show His forgiveness.

God made me imperfect, and He meant for me to be imperfect. I try hard to achieve perfection nonetheless. I face a *jihad* over these shortcomings every day. The concept of *jihad* is very misunderstood. For years, because of the mythology built around the Crusades wars, *jihad* has been understood in the West as "holy war." While the Qur'an instructs Muslims to face *jihads,* the Qur'an does not limit the meaning of *jihad*—struggle or striving—to physical struggles only, but also includes the struggle that is hardest of all—the inner struggle. For me, *jihad*

is a spiritual struggle or an internal struggle against temptations, against lying, against anything bad. *Jihad* and this idea that each person is struggling with himself or herself is actually very liberating. For most Muslims, *jihad* is seen as a challenge from God to improve oneself constantly—not as an excuse to make war. In fact, if *jihad* were supposed to refer to armed struggle, a variety of other words that mean violence or war in Arabic would have been better choices for the concept. *Jihad,* although it can mean an armed struggle, is just as much, or maybe even more, a personal struggle.

We are supposed to have our own black spots. They are meant, by God, to be there, just like the one in the Taj Mahal. We must engage in a *jihad* to overcome them. On our tour, the guide showed us the walls of the Taj Mahal. The walls consisted of panels of white marble, which master carvers had inlaid with precious and semiprecious stones to make a green and brown (among other colors) flower-and-vine design. Only a handful of the flowers and vines remained. Once adorned with the shiniest jewels from all over the subcontinent, the walls now had hollowed-out grooves where the jewels had once been. British military looted the Taj just before Indian independence. The once gushing and powerful river that banked at the foot of the Taj Mahal was nearly dried up. Though still amazing, the Taj Mahal, in its various states of disrepair, stands as much a monument to itself as to the beloved Mumtaz.

Once I realized that, in Islam, I was designed to make mistakes, I never felt better about myself. It's odd that learning I can expect to make mistakes would be comforting, but it was. I felt liberated from those mistakes and from the fear of making them. Islam gives me freedom to know that I am not perfect and, furthermore, that I do not have to be perfect. Only God is perfect, and I am not in competition with God. I am free to be second best.

When you are a perfectionist like I am, sometimes you have to resort to this reasoning to take your mind off of a mistake you have made. When things do not go the way I plan, I often say, "Well, I am not perfect, only God is. The fact that I did so well is quite impressive considering I was meant to fail!"

These *jihads* do not have to be serious ones either. I daily face a *jihad* not to overspend on my budget. Actually, I face a *jihad* simply to maintain a budget at all! We all have little *jihads,* and when you realize you are meant to have them, you can deal with them better.

One of the miracles of the Taj Mahal is that it is made of Indian marble, which has no pores. It cannot soak up any pollution or dirt. So while deprivation has affected neighborhoods like my dad's, the Taj Mahal has at least been spared most of the fallout from the elements. Italian marble, on the other hand, is porous and needs to be cleaned often. Large structures made of Italian marble are in a constant cycle of being cleansed. The Baha'i Temple built in Delhi in the mid-1990s, for instance, is made of Italian marble. Like the Taj, the temple is a massive white marble structure, but workers must clean it practically year-round to keep dirt and pollution from sullying it.

By contrast, the Indian marble of the Taj Mahal is perfect—incapable of soaking up dirt—whereas we humans are like the Italian marble of the Baha'i Temple—filled with numerous openings for pollution to seep in. But our deal with God is that He can cleanse us with His forgiveness. God says in the Qur'an, roughly: "I have forgiveness for you. In order to ask for my forgiveness, you need to make mistakes." So Islam gives you freedom to err and learn from your mistakes. God wants to forgive us. He loves us. We are His creations, after all. In the Qur'an, God told Muhammad to tell the early Muslims not to be scared of God: "[Muhammad,] Say [to the early Muslims]; 'If you love God, follow me, [and] God will love you and forgive you your sins; for God is much-forgiving, a dispenser of grace'" (3:31). A *hadith* attributed to Muhammad quotes him as saying, "All the children of Adam do mistakes, the best amongst them are those who repeatedly repent."

By design, I was late to the studio for my appearance on the ABC show *Politically Incorrect.* I knew that if I was late, I would be too busy in preparation for the show to become nervous. My brother, sister, and their friends all accompanied me, which was a little embarrassing since the bona fide celebrities in the green room with us had smaller entourages (if they had them at all). I was being shuttled between makeup and a preshow meeting with one of the producers. She was going to go over the proposed questions with me and my answers to them. She would later report these answers (along with the other guests' answers) to the host, Bill Maher, who would then decide at the taping which questions he would ask. Of the proposed questions, most had to do with Islam and Muslim women. The talk at the time was of the Taliban. It was October 2001. The attacks of 9/11 were still fresh in everyone's mind.

In all the excitement, I wasn't really nervous, mainly because I purposely showed up late. I knew, though, that I needed to be clear and not back down. Everybody I knew watched *Politically Incorrect,* and the show was going to air that same night. I think I realized I was doing a good job when another panelist, Malcolm McDowell, the British actor probably most famous for his role in the movie *A Clockwork Orange,* turned to me during a commercial break and said, "You really talk!" Before I knew it, the half-hour show was over. We posed for a photograph with Bill and were whisked away.

That night, as soon as the show aired, I received dozens of e-mails sent to my Web site. A lot of them were from young men who said basically that (1) I made good points, (2) Bill Maher is an idiot, and (3) I looked really cute. Of course, I was flattered and wrote back to some, defending Bill's open-mindedness in having a variety of guests on. I also received a few very critical e-mails from other Muslims.

My favorite response, though, was from my friend Tamir. Tamir is Arab, and his family has been American for generations. I only know him by e-mail and phone and his distinctive Texas drawl. One time, he introduced himself to me as the "Redneck Arab." That one surprised even me! He said he was really proud of how I spoke for Islam on *Politically Incorrect* and how graceful I was under the pressure of questioning. He reminded me of how in the Qur'an God responded to the angels' protesting His creation of humans by saying that the angels did not know what God knew. Tamir wrote that he felt God must have pointed down at me during the taping and said to the angels, "See, that's what I meant!" That even in a time when people of great faith were (and still are) challenged, when critics say that religion is the source of all conflict, and when the word *Islam* is often taken to mean terrorism, I could still show my pride in being a Muslim. That I did not hide it or act defensive but offered my knowledge and experience in an eloquent manner. The angels, when they protested the creation of human beings, *did not* actually know what God did—that we, as humans, are sometimes lucky enough to reach the highest of highs. Where Islam is judged daily on television talk shows, the outcome of *Politically Incorrect* or similar shows is very important. Doing a good job on *Politically Incorrect* could mean the difference between a soccer mom choosing to celebrate or revile Islam.

So many people today are disappointed with themselves. They always look at what they didn't achieve—how much they are being paid versus what someone else is being paid, what grades they received versus what grade the smartest person in the class received. They are constantly judging themselves.

What the Qur'an says is that God is the judge of us all and that He will not focus only on the end result but what you set out to do. Even in the section of the Qur'an on divorce, God reiterates forgiveness and the importance of good intentions:

"God will not take you to task for oaths which you may have uttered without thought, but will take you to task [only] for what your hearts have conceived [in earnest]: for God is much-forgiving, forbearing" (2:225). Little mistakes that are inadvertent will not be held against us. What we tried to do, despite the part of us that is hardwired to make mistakes, is what we will be judged on. What matters to God is our intentions: "But there is no blame on you if ye make a mistake therein: (what counts is) the intention of your hearts: and God is Oft-Returning, Most Merciful" (33:5).

God wants us, according to the Qur'an, to look at the world with curiosity and wonder. God, in the Qur'an, often lists His various creations as "signs" to us of His presence and design. The Qur'an encourages humans to take these signs and try to learn about them in a reasonable and logical way. Historically and even today, Islam, unlike other religions, has had little problem with science. Scientific discovery is religious discovery because God created the secrets of the universe for humans as signs:

> *Verily, in the creation of the heavens and of the earth, and the succession of night and day; in the ships that speed through the sea with what is useful to man: and in the waters which God sends down from the sky, giving life thereby to the earth after it had been lifeless, and causing all manner of living creatures to multiply thereon: and in the change of the winds, and the clouds that run their appointed courses between sky and earth: [in all this] there are messages indeed for people who use their reason.*
>
> (2:164)

In order to be curious enough to wonder about these signs and "use [our] reason," we must, as humans, be willing to take risks. To theorize why the sun rises every morning is to risk being wrong about that theory. When God asks us to explore the natural world for signs of Him, He also tacitly approves our employment of mistakes in the process. The end goal is to uncover these signs and learn about them, but we cannot do so without being wrong about them, at least initially.

Muhammad himself, the messenger of Islam, was a man who faced many challenges. If anyone had been doomed to fail, it would be someone like Muhammad—poor, orphaned, illiterate. But instead he became one of the world's greatest leaders. He was born into Arabia's most powerful tribe—but into the weakest subtribe. His father died while his mother was pregnant with him, and his mother died soon after his birth. An orphan, he was raised by his grandfather, who had to look out for Muhammad till his own death. Muhammad was also believed to be illiterate.

Born with no inheritance of his own, he had to work for his success. He went into caravan trading, and despite the disadvantages he faced, he prevailed to become recognized as a gifted businessman. The love of his life—Khadijah—who was not only his wife but also his boss (she owned the caravan trading business Muhammad worked for) was older than he and died when he was fifty years old. In the midst of all this personal upheaval, the Prophet Muhammad received the revelations that are preserved in the Qur'an.

Early Islam challenged everything that the tribal and feudal system of Arabia was based on, and Muhammad faced life-threatening danger because of the revelations he received. Contrary to popular belief, Muhammad never wanted to fight but was constantly forced to fight to save his own life and the lives of many of the early Muslims. For the first twelve years of his prophecy, he did not fight back but peacefully preached to his community, all the while suffering humiliating and mocking treatment from members of his own tribe and close relatives. He eventually moved from Mecca to Medina where he was able to forge alliances with other tribes and defend his community. In the end, even his most vocal enemies had succumbed to the beauty of Islam and the Qur'an.

He was constantly facing and overcoming *jihads* or struggles. Once he had unified all of Arabia in peace under monotheism (and in only twelve years), he became sick with the illness that led to his death. Single-handedly, he had brought Arabia out of a cruel and pagan tribalism to a smoothly functioning open society. From a penniless orphan, Muhammad became the messenger of the belief system that over a billion people subscribe to today. He was set for failure at his birth, but he never gave up on himself. Like all of us, he had his moments of difficulty, but his companions, particularly Khadijah and, after her death, his other wives, supported him. He picked up the pieces and kept moving. He is considered by Muslims to be as perfect a human as possible, and we all strive to emulate him. He was all he could be and more: a businessman, a philosopher, a theologian, a fierce warrior but also a fair and honest negotiator, a statesman in every sense of the word.

At any point in Muhammad's life, he could have easily, and with good reason, given up. But he kept on going, and the world would surely not be the place it is today if he had not.

I have to remember Muhammad's example at times in my own life. A few years ago I severely hurt my knee snowboarding and needed to use crutches or a wheelchair. My knee was rebuilt in surgery, but I couldn't put weight on it while it healed as I would risk falling. The night before my first day back to law school after the injury, I thought of myself having to climb the stairs leading to the entrance of the school, and I started crying. I wasn't sure if I could do it. Would I fall again? Would I be able to maneuver the crutches inside the door to where my wheelchair was being held for me at the coat check every morning? I was so scared, and I was mad too. I was mad at myself for being stupid enough to be injured. "You could have done this if you slipped in the bathroom," Dr. Hawkins, my surgeon tried to reassure me, but I was not placated.

I had gone snowboarding the first day of my spring break. I later thought that perhaps my greed to hit the slopes as soon as possible had resulted in my injury. When the ski patrol lifted me into the toboggan, I heard and felt my knee pop multiple times—as if fireworks were going off inside the joint. What was already torn up was tearing up even more. This knee was the same one I had hurt in high school. Stupidly, I wasn't wearing my knee brace, and now I was about to be dragged down the mountain. My brother, who was with me, looked at me apologetically. "It's popping!" I cried. He blurted out, "Oh God," and then regained his composure and tried to look like everything was fine. As much as my brother wanted to take my injury away, he couldn't. No one can do these things for you. Some things you just have to do yourself. Now just climbing the stairs to my law school seemed insurmountable.

My mom had told me once that Muslims believe God only tests people whom He thinks can pass that particular test. We are all graded on a curve built specially for us. "Good foot to heaven, bad foot to hell," I said to myself. The mantra of those on crutches reminded me to use my uninjured leg first going up the stairs and then my injured leg going down. The first morning I returned to classes, I simply climbed the stairs. It was the hardest and easiest moment of my life. I wasn't unifying the tribes of Arabia like Muhammad had, but I had met my own *jihad* for that period of my life.

Whenever I straighten my knee these days, especially if I have been exercising hard, my knee snaps, crackles, and pops—probably scar tissue that I never properly rehabilitated through physical therapy. Dr. Hawkins says he can go in again and remove the source of the popping, but I am not so eager to experience another surgery. He has also cleared me for snow sports, but I am not ready yet.

Years before these injuries, back in boarding school when my body was still unscarred and my joints didn't pop continually like the train to the Taj Mahal, I attended required daily chapel every morning. At the end of one particular chapel, we were going to sing "Morning Has Broken" by Cat Stevens as our closing hymn. I faced the usual dilemma I did as an eighth grader—do I sing the hymn or not? I faced this tiny *jihad* every morning we had chapel. I knew that taking communion was insensitive to those who believed in communion. Would singing an Episcopalian hymn be similarly offensive? Furthermore, was it an offense to my own religion to sing such a hymn?

The service was usually secular—we opened chapel as often with prayers from the Native American tradition as we did with Episcopalian prayers—but the hymns were clearly Christian in nature. Was it wrong for me to sing them?

When I had first started attending boarding school, the chapel fascinated me. It was always dark. In fact, I don't think there were any lights in the main section. The chairs were all wooden and attached to one another in long rows. So if you pulled one chair, you dragged the entire row with it. The students, on their own, mainly sat according to age, with the younger students in the front and the older ones in the back. The seats had a wooden frame but with a patch of wicker as the seat. Some wicker was practically torn off from use. The stone aisle between the two main sections was made in a black-and-white checkerboard pattern. The blue and red stained-glass windows presided over the chapel and had the only notable color in the entire place except for the red velvet banner hanging behind the cross at the front of the chapel. Faint gold steel grates dotted the floor, leading to what seemed like a bottomless pit but must have actually been the basement.

On a pillar hung a foot-wide wooden board that listed the number of the hymn we were going to sing. The numbers and letters for the board were like the kind used for movie-theater marquees but not nearly as large. The hymnal and one copy of the *Book of Common Prayer* were tucked into little wooden racks built into the back of each chair—sort of like a built-in bookshelf. You would take the hymnal from the back of the chair in front of you. If you were bored during the chapel talk—a speech given by a member of the community and usually not religious—you could look up that day's hymn during the talk.

The headmaster, Mr. Polk, usually gave one chapel talk a week. He would stand high up in the pulpit, which had its own little light and light switch. He always, and especially in these talks, exuded a moral clarity and stillness that I have yet to see in anyone else. Both students and teachers had one designated day a week for talks. Mr. Connor, the Spanish teacher, gave the funniest chapel talks, and my sister even gave one when I was in ninth grade—on our parents' arranged marriage. She was a senior then. At the end of the first chapel talk I heard, on my very first day of school, I began to applaud. I had always applauded at the end of speeches. This time, everyone near me stared in my direction in horror. Apparently, one does not applaud in chapel! I was horrified myself at all the attention I had drawn and immediately stopped clapping.

I usually had only a few seconds to decide if I would sing the hymn or not. My decision this particular morning was made more urgent by the fact that "Morning Has Broken" is actually a pretty short hymn—it would be finished before I could decide! We usually sang longer ones. One really long one called "God Be with You Till We Meet Again" we always sang before vacations and long weekends. My favorite hymn, which I always sang, was "Jerusalem," based on the William Blake poem of the same name. It was so dramatic:

And was Jerusalem builded here
Among these dark satanic mills?
Bring me my bow of burning gold!
Bring me my arrows of desire!
... I will not cease from mental fight,
Nor shall my sword sleep in my hand
Till we have built Jerusalem
In England's green and pleasant land.

But we were not singing "Jerusalem." We were singing Cat Stevens! I made the split-second decision to sing. After all, Cat Stevens's now being Muslim meant it was probably okay for me to sing his song:

Mine is the sunlight, mine is the morning
Born of the one light, Eden saw play.
Praise with elation, praise every morning
God's re-creation of the new day.

Cat Stevens converted to Islam in 1977 and later changed his name to Yusuf Islam. A lot of his fans surely felt betrayed, especially when he decided not to perform music for a period based upon the bad advice of an extremely conservative Muslim. In his fervor to convert, he felt that he needed to dump most vestiges of his previous rock-star lifestyle, including performing music. When he first began the conversion process, he says, what drew him was not other Muslims but the Qur'an. On his Web site, www.mountainoflight.co.uk, he later wrote: "I read the Qur'an first [before meeting other Muslims] and realized that no person is perfect. Islam is perfect."

Like for me, Yusuf found this aspect of Islam liberating. He realized that following the will of God was a feat that God's other major creation—the angels—couldn't perform. This realization made him want to be a Muslim desperately: "I realized that everything belongs to God." He also lost what he calls "pride." "I had thought the reason I was here was because of my own greatness," he writes on his Web site. "But I realized that I did not create myself, and the whole purpose of my being here was to submit to the teaching that has been perfected by the religion we know as Al-Islam. At this point I started discovering my faith. I felt I was a Muslim." Eventually, Yusuf began singing and recording music again but only songs related to Islam. He donates the proceeds from royalties on his more popular music to Islamic causes.

"Morning Has Broken," which was recorded well before Cat Stevens converted, expresses a principle of Islam that later motivated his conversion. "Morning Has Broken" describes the morning as the perfect manifestation of God's creation. The morning on earth, with its sunlight and glory, was recreated every day by God. Stevens exhorted us in "Morning Has Broken" to praise God's perfection. Despite our flaws, we have a glimpse of Eden every morning. I didn't know it at the time, but in singing "Morning Has Broken," I was espousing the very principle of Islam that I would later take comfort in after my knee injury.

Shah Jehan, who lived centuries before Cat Stevens, also had this understanding. He knew that as perfect as the Taj Mahal was, it would never be as perfect as God. He deliberately added the black spot to emphasize the point. In fact, many say the Taj Mahal is most beautiful in the morning—when the sun rises and washes the marble in light, the white gleaming in shades of pink and orange—"God's re-creation of the new day" at the Taj Mahal, as Cat Stevens would put it.

The Taj Mahal's black spot would be accompanied by other problems that affected its near-perfection. Shah Jehan's son, Aurangzeb, became the Mughal emperor after his father by killing his own brother. The brother was his nearest competitor for the throne, and Shah Jehan had already selected the brother as his successor. Aurangzeb had to kill him if he wanted to be emperor. He didn't kill his father, despite the father's previous endorsement of his brother, although in those days such a deed clearly warranted execution. Aurangzeb instead let Shah Jehan live under house arrest in a castle fort opposite the Taj where he could still cherish the memory of his beloved wife. His daughter lived with him and took care of him. Romantic historians say that the room in which Shah Jehan lived out the rest of his days had a small hole for a window with a perfect view of the Taj, which he gazed at night and day till his death.

When Shah Jehan finally died, Aurangzeb placed his father's tomb alongside his mother's. Doing so was cheaper than building a separate sanctuary for him, which according to lore Shah Jehan had planned as a black marble version of the Taj Mahal across the river from the white one, although no historical record of such a plan exists. Until Shah Jehan's death, if one had divided the Taj Mahal down the middle—from the center of the point of the dome, down to the floor, the two portions would perfectly mirror each other—like the way the reflecting pool in front of the Taj Mahal reflects the structure every day. With the looting and the general disrepair of the surrounding area, this great achievement of Shah Jehan's reign has been further compromised. Even the great Shah Jehan or his fierce son Aurangzeb could not have foreseen the Taj Mahal's future. The Qur'an would say of them: "Thou seest the mountains and thinkest them firmly fixed: but they shall pass away as the clouds pass away: (such is) the artistry of God, who disposes of all things in perfect order: for He is well acquainted with all that ye do" (27:88). Even despite the trials the Taj Mahal has gone through, it is still a gleaming white palace with no match to this day. No one would say that the "mistakes" that have affected it have taken away from its overall majesty. Only God can be perfect. We, fortunately, do not have to be.

Chapter 5

The Diversity of Islam

O mankind! We created you from
a single pair of a male and a female,
And made you into nations and tribes,
that ye may know each other
(Not that ye may despise each other).

Qur'an 49:13

"You must come to the Eid prayers with us," said Zainab Auntie to me over the phone the night we arrived in Chicago. Zainab is not actually my aunt; she is the wife of an old family friend, but in South Asian culture, one generally calls family friends who are the same ages as one's parents "aunt" and "uncle." Zainab's husband and my father had done their medical residencies in Chicago when they both had first moved to America from Pakistan and India respectively. While her husband had stayed on in Chicago to practice radiology, my father had moved his family to Pueblo.

My mom, dad, sister, brother, and I met in Chicago for a reunion of sorts to coincide with Eid al-Udha, a Muslim holiday that commemorates Abraham's near-sacrifice of Ishmael, exemplifying Abraham's devotion to the one God. We arranged to meet my auntie and uncle the next morning.

On the first morning of the three-day holiday, my family walked over to McCormick Place, a convention center, where Eid prayers were being held. The huge turnout at every mosque each year finally spurred the community to rent two gigantic halls in the downtown convention center. Finding a spot in the women's section, my mother and I sat down and waited for the service to begin.

We were early (for once) and had the hall mostly to ourselves. Long strips of thin, white paper, about four feet wide, were spread in diagonal lines all over the hall. The stone floor was cold and hard to the touch. The strips of paper were being used in place of prayer rugs. If everyone who came to the prayer brought a rug, the hall would be filled in minutes with small Persian carpets, probably mostly with pictures of the Kaaba in Mecca on them. The paper saved space and would be easy to clean up. It sat neat and, amazingly, clean.

As the time for the prayers came closer, the hall swelled with activity. Massive numbers of people streamed in. For any other occasion and perhaps with a non-Muslim crowd, the prayer paper would have been a rumpled mess before the start of the ceremony, but by the time I noticed the groups of beautiful and statuesque African American women who came to pray, the paper was still intact, just waiting for some Muslims to pray on it. The women had on long dresses made of orange, black, and green African Kente cloths with matching headdresses. I had never seen such voluptuous and colorfully dressed women before in my life.

My mom and I smiled at each other, both of us clearly surprised at how our normally festive *shalwar-kameezes*—our Pakistani ethnic outfits, which consisted of colorful, harem-type pants and a long tunic—were being upstaged! We also saw more South Asians and other Asian races. But nothing quite struck me as much as the African American women, who I can still clearly see today, walking through my memories like they have found a home there.

My grandmother—the one who chose the name "Asma" for me—had gone on the *hajj* pilgrimage for the first time in 1970. There, she too was awed by the African female delegation, just as I would be years later. The *hajj* is a pillar of Islam, required of every healthy and financially able Muslim. One of the points of *hajj* is to meet Muslims from other countries and backgrounds, or at least see them. The *hajj* is a multiday event, ending in a mass prayer ceremony on Mount Arafat near Mecca, where the Prophet Muhammad gave his last sermon before dying. The entire group of pilgrims prays to God all day long for His mercy and forgiveness for their sins. Pilgrims are required to wear white clothing without seams, assuring that no distinction can be seen among them.

The gathering is a preview of Judgment Day as described in the Qur'an, when all people will assemble on Mount Arafat to present themselves to God. As humans, we see the differences of skin color. The seamless, white clothing is meant to symbolize how God sees us, no differences of any kind—color, economic, or otherwise. The Qur'an describes diversity as part of God's plan: "O mankind! We created you from a single pair of a male and a female, and made you into nations and tribes, that ye may know each other (Not that ye may despise each other)" (49:13). God made us diverse so that we could enjoy our diversity. For God, the only difference between us is our piety.

Like my grandmother, the African American civil rights leader Malcolm X had an eye-opening experience at *hajj* too. His astonishment, however, came from seeing the white Muslims, not the African ones. In a famous letter to his wife, Betty, he wrote about his experiences at the *hajj*, like sharing his sleeping tent and also water with white Muslims: "There were tens of thousands of pilgrims, from all over the world. The Qur'an predicted this diversity: 'And among His wonders is the creation of the heavens and the earth, and the diversity of your tongues and colors: for in this, behold, there are messages indeed for all who are possessed of [innate] knowledge!' (30:22). They were of all colors, from blue-eyed blondes to black-skinned Africans. But we were all participating in the same ritual, displaying a spirit of unity and brotherhood that my experiences in America had led me to believe never could exist between the white and non-white."

At *hajj*, he realized that Islam was not just the message of one race but also a universal message for all. He even felt that if Americans understood Islam better, then they would see how God intended the various ethnicities and diversity of Muslims to be a blessing. His life was changed forever by the *hajj*, as it is for many Muslims. His death soon after cut short his attempt to "know," as the Qur'an puts it, diversity, but, like Malcolm X, I am proud to practice a religion that is also practiced by members of every race in the world.

Arabs, Persians, black Africans, Turkish people, Indian and South Asian, Malaysian, Mongolian, Russian, Eastern European, white Americans, and Chinese people all practice Islam and all over the world. Growing up in small-town Pueblo, Colorado, I did not really fully appreciate just how diverse Islam is until I was in college. I was a faithful member of Al-Muslimat, the Wellesley Muslim students group. My sophomore year, at our opening meeting of the year with all the new members, I saw an Asian girl. I thought that she must either be a convert or in the wrong meeting, but she seemed very interested. She was dressed in black baggy pants and a snug red T-shirt. She looked like she might ride a skateboard and had medium-length hair cut all to the same length. She was thin and tall with a dark complexion.

After the meeting, the group socialized. Though a little timid, I asked her if her family was Muslim.

"Yeah, of course," she said.

"Soooooo," I said tentatively, "you must be South Asian." Clearly she wasn't, but she took my olive branch and ran with it. I was probably not the first Muslim to wonder where she came from.

"I'm Vietnamese," she said. "A Cham Muslim. My family has been Muslim for generations."

"Wow!" I said. I had liked her from the moment I saw her—she had attitude, but she wasn't mean—but I liked finding out that she was Muslim too. It was exciting. I don't remember her name, but I saw her at Muslim events often. Since then, I've met Muslims from all kinds of backgrounds, but it took the Cham skateboarder girl to make me open my eyes to Islam's diversity. I later learned that Cham Muslims are mainly Muslims of Cambodian and Vietnamese background. Faced with years of war in their countries, they have not only suffered as the rest of their countries have but also as religious minorities in their own countries at the hands of oppressive governments.

As I would discover, Islam was diverse from the start and reached out to those who had traditionally been left out. Some of the first converts, who converted after Muhammad's early prophecy, were from the clans and subtribes at the bottom of the societal hierarchy. Slaves and women in addition to other disadvantaged members of the premonotheistic, pagan society of Arabia also converted. A large number of the first Muslims came from the oppressed classes of Arabia. Under Islam, God turned away no one because all were equal in God's eyes. Being the chief of your tribe versus being an orphan, although a night-and-day difference to the pagan, pre-Islamic Arabs, meant (and means) nothing to God.

When people find out I am Muslim, they often say something like, "I've never met an Arab before," or, "Being Arab American must be hard these days." While I don't doubt that either of these statements is true or sincere, they don't apply to me. I am not Arab. Though the religion of Islam began in Arabia, it has spread far beyond that land. In fact, the vast majority of Muslims today are not Arab. The Arab Muslim population in the entire world is only about 20 percent of the world Islamic community. The percentage of Arab Muslims among American Muslims is also around 20 percent. A lot of non-Muslims mistakenly assume the contrary—that all Muslims are Arab and that all Arabs are Muslim.

Actually, most Muslims are Asian—South Asian, Indonesian, Malaysian, and Chinese. The two countries with the largest Muslim populations are Indonesia and India, two countries far from the Arab world (although I've heard people comment on the troubles between India and Pakistan as part of the "Middle East conflict"). Similarly, many Arabs are not Muslim. Arabs can be Christian, Zoroastrian, Druze, Baha'i or many other religions. Islam itself is a religion, not an ethnicity. To say that one is a Muslim is simply to say that one believes in one God and that Muhammad is His messenger. This declaration does not indicate any ethnicity.

What brought Islam to so many people? Revisionist historians would like us to believe that ruthless Islamic tyrants forced their conquests to convert or die. Though certainly some Muslim leaders encouraged conversion, the hype is misplaced. Muslims are not required to engage in missionary work aimed at converting non-Muslims. Some Muslims, especially those who practice the Wahhabi (also called Salafi) interpretations, do actively preach to other Muslims. However, in my experience even the Wahhabis, for all their radicalism, do not try to convert non-Muslims. The Qur'an is clear: "Let there be no compulsion in religion" (2:256).

Religion cannot be forced on anyone—whether it is Islam or Christianity or any other religion. The four caliphs (also called the *Rashidun* or rightly guided caliphs) who led the Muslim community after Muhammad's death had a multi-faith perspective ahead of their time. Following Muhammad's example, they did not seek the conversion of non-Muslims.

Both sides of my family probably became Muslim because of the famous Silk Road. Until the thirteenth century, the only way for Europe and the West to obtain the much-sought-after silk was through a long journey of caravans and trading. At its peak, the Silk Road connected Hong Kong through China to Uzbekistan and eventually even to Venice by sea. The Muslims who participated in the Silk Road trading were like Muhammad—caravan traders and businessmen. The exchange of silk, of course, brought not only silk goods but also an exchange of ideas. The Silk Road was like the Internet today. People of different cities and towns met along the paths of the route and trading posts, where they exchanged ideas and dialogue. Islam spread to many peoples through the Silk Road, especially to the people of China.

When Vasco da Gama found a sea route to China, the Silk Road lost its importance. But Islam still finds an audience in a variety of ways. When I was in middle school, I remember reading a storybook about Tariq bin Ziyad, the general of the Muslims who conquered Spain in 711. These Muslims later became known as the Moors after the reconquest of Spain by Christians. The Moorish influence on Spain was tremendous—everything from language and architecture to dance and art was affected by Islam and Arab culture. When I learned Spanish in high school, I found out that the Spanish term for the word "there" is *alla* (pronounced "uh-ya"). Because when the Spanish Muslims pointed to the sky, they said "Allah," the locals assumed they were pointing out that something was *there* in the sky. Hence the adaptation of the Arabic word for God to the Spanish word for "there."

The Moors seemed pretty interesting to me. So I was disappointed to find out that no Muslims live in Spain anymore. Though the Moors intermarried with the Spaniards, the practice of Islam in Spain ended. Most of the Spanish Muslims were thrown out by royal edict along with Spanish Jews. The remaining non-Christians converted to Christianity to save their lives. The Great Mosque of Cordoba, the masterpiece of Moorish Islamic architecture and design, with its striped red and white archways, was converted into a church. The Christian extremist movement ended a great period of multicultural and multifaith cooperation and achievement in world and Spanish history.

Years later, when I began researching Islam in America, I found out that about 40,000 American Muslims are of Latin descent. Some of them feel they are returning to their Moorish roots, while others simply found the message of Islam attractive. One man told me that he first became intrigued about Islam because of the music and lyrics of Muslim rappers like Brand Nubian, Poor Righteous Teachers, Erik B. & Rakim. Later on, when he seriously began studying Islam, he realized that some of the rappers' lyrics were not actually very Islamic and that they followed an unrecognized rogue branch of Islam, whose followers call themselves "5 Percenters." He also learned of famous rappers who follow mainstream Islam, including Mos Def, Q-Tip, and Erik Schrody of Everlast. He said that Islam was the first religion that made sense to him and made him feel equal to other ethnicities.

Having moved from Venezuela to America when he was twelve, he wanted to be part of a community that was global and not focused on what his differences were. Now, he told me, he attends a majority Pakistani American mosque in Miami and wears the male version of the *shalwar-kameez* to the Friday prayers many Muslims attend.

Muslims can be white too. Eastern European Muslims from Albania, Russia, and other European countries are numerous, but because of their color, they are rarely noticed as part of the Islamic religion. I've also met a large number of white, American women who have converted to Islam after falling in love with a Muslim man in grad school or college. The Silk Road is still traveled today, but it comes in the form of Muslim rap songs and master's degrees.

In the United States, a little less than half of the Muslim community is African American. Like Malcolm X, most African American Muslims converted during the civil rights era, and many of their family members have been Muslim ever since. Even today, African American Muslims convert to Islam because, like many Latin American Muslims, they want to return to the religion of their ancestors. Many slaves brought to America were Muslim and lost touch with Islam over time. Now many African Americans who convert feel they are regaining what was taken from them.

Although historical records are incomplete, scholars believe that of the over ten million Africans brought to America as slaves, as many as 30 percent were Muslim. Over 60 percent of Africa today is Muslim, so the number of African slaves who were Muslim may be even higher.

Several years ago, when I was writing my senior thesis on Islam in America, I visited Auntie Zainab in Chicago. The lack of information and research on American Muslims forced me out "into the field," as we would now say. I had traveled all over the country interviewing American Muslims and had stopped in Chicago to interview Zainab about the organization she had cofounded. Whizzing through Chicago from the airport, Zainab began telling me about Apna Ghar. Years earlier, Zainab, along with other South Asian women, opened a shelter to support women of South Asian background who had been abused by their husbands.

"So you mean the shelter helps non-Muslim women too?" I asked her, surprised. Although Zainab had never said the shelter only helped Muslim women, for some reason, perhaps out of a focus on my thesis topic, I assumed it did.

"Of course it does," she said. As it turned out, the shelter would help any woman with special cultural needs, even if she wasn't South Asian.

At Apna Ghar, which means "My House" in Hindi or Urdu, the main languages of North India, Pakistan, and Iran, I met an immigrant Iranian Muslim woman and her daughter. She had been abused by her husband. She didn't want to return to Iran, but she didn't know how she could stay in America. I also saw plenty of non-Muslim women there, both residents and volunteers. Of course, I ended up including Apna Ghar in my thesis, precisely as proof of American Muslims' assimilation into American society and as a reflection of American values. The diversity discussed in the Qur'an—both regarding ethnicity and also understanding of other religions—is certainly not lost on American Muslims.

As a Muslim, I believe that each person has a different path to God. One person may find God in Buddhism, another may take comfort in the abstract idea of a higher being, and still others formally practice Christianity daily and so on. The

Qur'an says that for each group of people, God has sent a messenger with a law to follow: "Unto every one of you have We appointed a [different] law and way of life. And if God had so willed, He could surely have made you all one single community: but [He willed it otherwise] in order to test you by means of what He has vouchsafed unto you" (5:48). The Qur'an also says that on Judgment Day, when we gather at Mount Arafat, we will each be judged according to the belief system we chose. So any Muslim who says, "You will not be saved because you are not Muslim" does not understand the Qur'an.

Ever since I published my first book, evangelical Christians often have exhorted me to find my salvation in Jesus Christ. "Jesus said, 'I am the way, the truth, and the light,'" they write to me, quoting a famous New Testament verse. The pages that follow are usually attacks on Islam, false (and, frankly, very offensive) claims that Muhammad raped little girls, and so on. I am not really sure how this pitch is meant to convert me to Christianity. I certainly would never describe Islam to someone by attacking Jesus in such a manner.

I'll never forget a Christian classmate of mine at Wellesley telling me that she would miss me in heaven.

"Why?" I asked her.

"Well, because," she paused, "only Christians can be saved, and you're not Christian." I was stunned! How could my friend honestly believe in a religion that would exclude someone she liked simply because that friend was not a member of that religion?

I grew up knowing that Muslims accepted all religions that were moral and good. The discussion with my "friend" made me realize that this acceptance was not universal among religions. In Islam, you do not have to be Muslim to be "saved." The Qur'an says: "Verily, those who have attained to faith [in this divine writ], as well as those who follow the Jewish faith, and the Christians, and the Sabians—all who believe in God and the Last Day and do righteous deeds—shall have their reward with their Sustainer; and no fear need they have, and neither shall they grieve" (2:62).* Besides the ethnic diversity of Islam, Islam recognizes that other religions exist and that God accepts those religions so long as they are based on justice and morality.

Muslims don't have a problem with Christians or Jews or other religions—at least not by Qur'anic standards. Christians and Jews are called the "People of the Book" in the Qur'an because of the great similarities between the Torah, the Gospels, and the Qur'an. Muslims actually call Christians and Jews "Brothers and Sisters of the Book." The title is a homage to Ishmael and Isaac. Just as they were brothers, Muslims, who are the descendants of Ishmael, feel that Christians and Jews, who are the descendants of Isaac, are their siblings too.

When I was growing up, my parents and other Muslims always told me to learn about Judaism and Christianity. "They

---

* This particular verse, which posits that, under Islam, all good people will be rewarded in the afterlife, is repeated two more times in the Qur'an at 5:69 and 22:17.

In addition, the Sabians were a small religious community present in Arabia at the time of the Prophet Muhammad's life. They may have been Christian followers of Saint John and possibly practiced a religion similar to Zoroastrianism. Not a lot is known about them, and their exact identity is unclear. Although they do not exist anymore today, they are mentioned in the Qur'an several times.

are our brothers," people like my grandfather would say to me. No Muslim ever said anything negative about Christians or Jews to me, even today. Even in political disagreements, a Muslim is (and should be) careful to distinguish between the common upheavals of politics and the sincerity of continuous religious belief.

The Qur'an never said that Islam negates the teachings of previous religions. In fact, to the contrary, Muslims are instructed by the Qur'an to read the Torah and the Gospels, as these books are part of God's continuing revelation to humankind. The Qur'an even instructs Muslims not to argue with non-Muslims: "And dispute ye not with the People of the Book . . . unless it be with those of them who inflict wrong (and injury): But say, 'We believe in the Revelation which has come down to us and in that which came down to you; our God and your God is One; and it is to Him we bow (in Islam)'" (29:46). Some Islamic scholars, including Karen Armstrong, feel that if Muhammad had known about other religions like those practiced by Buddhists, Hindus, Native Americans, and Australian Aborigines, he would have celebrated these religions too.

The view of the Qur'an is that all religions that strive for justice, stand for equality, and ban the worship of idols are derived from one source—God. Although Islam is the final and complete revelation, Islam is a stone in the road to God on which all religions focusing on God are present. The Qur'an says that for each people, a new prophet came, concluding with Muhammad. God's message was revealed slowly and throughout time by different messengers and in a manner suited to the time period.

Islam encourages diversity—outside Islam and within it. The Prophet Muhammad praised diversity. A famous *hadith* attributed to him is: "Difference of opinion is a mercy for my community." Although some scholars now feel this hadith may not have been said by Muhammad, the meaning behind it was commonly understood and promoted by Muhammad's companions. Furthermore, Muhammad's life is filled with examples of his reaching out to non-Muslims in a spirit of companionship. For instance, Muhammad keenly wanted to show solidarity with Jews. He even instructed the early Muslims to fast on the Jewish Day of Atonement, a fast he himself observed as well.

Muhammad never asked Jews or Christians to convert, because, according to the Qur'an, they had received revelations of their own suited to them. Any person can convert if he wants to, but to expect people to convert as the only source for salvation is considered absurd in Islam. When Muhammad first received the revelations, the two people who strengthened his resolve despite his own personal doubt over what the revelation he received meant, were his wife Khadijah and her relative Waraqa ibn Nawfal, who was a devout Christian. He shared with Muhammad what he knew of Jesus' prophecy and agreed with Khadijah that Muhammad must be a prophet, too, based on what had happened to him. But Waraqa never converted nor did Muhammad expect him to. He had already received the revelation best suited to him.

Under Islamic law, religious minorities are to be protected as Muhammad stipulated in the Medina Constitution and in his own leadership of the Medina community. The Medina Constitution specifically said that the majority Muslim community of Medina would look after the pagan and Jewish communities of Medina and that they had formed a society together. Of course, at times, individual Muslim rulers have persecuted religious minorities. But if we judged Islam by the failure of individual Muslims to follow it, we would be ignoring the fact that as long as 1,400 years ago, Muhammad established a legal framework for protecting people who were different. Similarly, pagan religions that date from before the founding of Islam are still practiced in some Islamic countries including Iran, Iraq, and Pakistan, and were not forcibly eradicated.

Under Islamic law, religious minorities could be made *dhimmis,* which in Arabic roughly means "protected subjects." Once given this status, a religious minority could not be attacked by a Muslim. As the Muslim empire grew, so did the number of *dhimmis*. Non-Muslims living in Muslim-controlled land were required by Islam to be protected by the Muslim stewards. Muslims were proud to protect these minorities and even avenged wrongs done to *dhimmis*. Early Islamic rulers felt almost a macho sense of pride in protecting their *dhimmis* who were not required to fight on behalf of the state.

One of the core principles of Islam is justice. In Islam, everyone deserves justice, and everyone must reasonably enforce it. To promote the cause of justice included protecting the *dhimmis*. Protecting *dhimmis* inherently meant practicing Islam. These *dhimmis* paid a tax to their protectors in exchange for the protection they received and practiced their own religions freely.*

A Christian king who assisted the early Muslims may have actually influenced the *dhimmi* concept. Before the exodus of the young and new Muslim community to Medina, the early Muslims lived in Mecca. Mecca was a very different place than it is today. A city that had become prosperous from trade, its inhabitants were averse to the changes Islam proposed—the equality of all people, poor or rich, the freeing of slaves, and so on.

The *hajj* pilgrimage, which predates Islam, was the major Arabian trade destination in pre-Islamic times. The pagans of pre-Islamic days came to Mecca to worship the idols then housed inside the Kaaba. The pilgrimage also gave the pre-Islamic Arabs the opportunity to meet yearly, in a sort of convention, and to trade or sell their various goods among themselves. Pagan Meccans were worried that Muhammad's new religion, with its anti-idol stance, would destroy this major source of trade revenue. Believing that Muslims were a threat to their livelihood, many pagan Meccans dedicated their time to defeating Muhammad's new movement. The converts were humiliated on a daily basis by their own family members and neighbors. They were spat on and beaten in a communitywide effort to cause them to abandon this new faith. Muhammad

* Incidentally, *dhimmis* who had their own means of defense or who joined the Muslim army did not have to pay the tax. In addition, women and children were not taxed either because they were not required to defend the community.

was spared most of this abuse, as his grandfather, who had not converted, was a local leader. Any attack on Muhammad, even if he was the founder of this new religion, would require that his grandfather avenge the attack, simply to preserve family honor. Arabia was caught in this tribalism, and Islam would eventually liberate Arabia. But nothing was going to change overnight (although, in only ten short years, Muhammad united Arabia in peaceful monotheism).

In the early history of Islam, a Christian ruler was the first to help Muslims. To spare themselves the constant discrimination, several of the early Muslims sought and were given permission by Muhammad to flee to neighboring Abyssinia, where the Christian king Negus promised to give them shelter. Reportedly, when he received word from Mecca that several Meccans were coming to Abyssinia to kidnap and torture those who had fled, he asked the Muslims to make their case to him. The early Muslims recited to him the portion of the Qur'an about the Virgin Mary and the birth of Jesus. Moved by these passages and respecting their apparent religious devotion, as he was a religious man himself, Negus protected the Muslims. Muhammad himself remained in Arabia but eventually fled to Medina where the entire Muslim community, including the group who found sanctuary in Christian Abyssinia, was reunited. In Medina, the *dhimmi* concept began to be developed. Muslims themselves had been *dhimmis* of the Christian king of Abyssinia, having firsthand knowledge of religious oppression.

The concept of *dhimmis* is frequently misquoted and misconstrued. In fact, to some, simply saying the word *dhimmi* has now become an indictment of Islam, as if it requires no explanation. People try to paint the *dhimmi* distinction as tantamount to discrimination or segregation and an obvious marker of Islam's failure to deal well with minorities. Besides the fact that history reveals otherwise, these critics fail to recognize that regardless of how the *dhimmi* concept played out, it was revolutionary for its time and was also meant to be a good thing. The idea of protecting and supporting those worse off than you is an Islamic one, regardless of how critics mischaracterize it.

I often feel that arcane items like the *dhimmi* concept are used to misrepresent and prejudice others against my religion. I'd never even heard of *dhimmis* until I went to college and studied Islam in a textbook. My living Islam did not include any references to *dhimmis*. I once attended a conference consisting of a who's who of American life—everyone from former U.S. presidents and politicians to dot-com entrepreneurs. Feigning liberal political views, in a self-congratulatory setting, these elitists would repeat the deepest and most troubling stereotypes about Muslims—and to my own face. I wanted to say, "don't you think that if Islam really stood for that, I wouldn't practice Islam?"

At this conference a former astronaut confronted me in a roomful of other participants. He claimed that Muslims are eager to bring back the era of *dhimmi* standards when Christians were made to ride on mules at least a certain distance behind the superior-feeling Muslims. I tried to answer his question, but he actually walked out of the room. I asked him to stay and listen to what I had to say, but he ignored me and, in front of the entire room full of conference attendees, walked out on me. I couldn't believe it. This man had been to outer space, but he couldn't listen to a two-minute answer? He had rocketed through our atmosphere, but he couldn't entertain changing or discussing his views on Islam? He was so bound to his prejudice that he didn't even give me the courtesy of listening.

Another source of misrepresentation is the number of Qur'anic passages warning Muslims not to become friends with Jews, Christians, or other non-Muslims: "O ye who believe! Take not the Jews and the Christians for your friends and protectors" (5:51). A variety of people quoted these passages in the months after the tragic 9/11 attacks as some kind of cryptic proof that Islamic terrorism was divinely ordained in the Muslim mind. Ignoring the fact that unprovoked, non-military killing is forbidden in Islam, these passages are not a blanket approval for Muslims to treat Jews or Christians badly. They were written for a particular situation Muhammad was facing and were meant only to apply to that time. As with passages in the Torah and Bible, the reader of the Qur'an should realize that each verse has a historical context. Chapters eight and nine of the Qur'an, for instance, specifically deal with the war being waged in Arabia against the early Muslims. To take a passage with such historical context and blindly apply it to all times is not a proper interpretation of Islam. Furthermore, the use of the term "friends" is a mistranslation. Each word in Arabic has a variety of meanings. Here, the word for "friend" should have been translated into "master" or "protector," which would make more sense given the historical context. But because Yusuf Ali, a Pakistani who did the most popular translation of the Qur'an into English, had not grown up speaking Arabic, he did not recognize that "friend" was not the appropriate choice for the translation of these verses.

The Jewish tribes the Qur'an warned Muhammad against were among those groups plotting the demise of Muhammad and Islam. The pagans of Mecca made several assassination attempts on Muhammad while he lived in Medina, and they were assisted by Jewish, Christian, and pagan tribes that neighbored Medina and who had previously (and apparently falsely) made their peace with Muhammad. Betraying their truce with Muhammad, they provided information about Muhammad's whereabouts to the pagan Meccans and gave them shelter. Their behavior had nothing necessarily to do with their religion, but, for purposes of identifying them, the Qur'an called them Jewish or Christian or pagan as the case may have been.

The Qur'an, like all the great sacred texts, is not supposed to be read out of context. Anyone who cites the passages about not taking Christians and Jews as friends is not only following an erroneous translation but also failing to account for the historical context of those verses. They are ignoring (or blatantly defying) the rest of the Qur'an, which celebrates religious and ethnic diversity. In fact, the very next verse for most of those passages exhorts Muslims to forgive those who have wronged you, regardless of his or her religion.

The Qur'an describes God as the "light" that guides all humankind who do right, not just Muslims. The most important belief in Islam is that God is one. He has no partners or coequals or even competitors. As a result, all those who believe in God must believe in the same God. The Qur'an stands for the principle that all just religions come from God. When God's light shines, it illuminates all who see the light, according to the Qur'an (24:35). Nowhere does it say only Muslims will see this light or only the Muslims among those who see the light will be rewarded.

Shortly after we moved to Pueblo from Chicago, my father asked my mother when she was going to become a U.S. citizen. She told him she didn't know. She had done all the necessary groundwork—filing the citizenship paperwork and so on. All that was left was to apply. My father had become a citizen as soon as he could. He held no devotion to the country of his birth, India, for he had been expelled from it. As a young adult refugee in Pakistan, the country of his youth and midtwenties, he realized that it held no opportunity for him, and he questioned Pakistani attitudes on fairness and human rights. He couldn't wait to be an American.

Since the days he had read books at the American library in Lahore, Pakistan, he had known he loved America. It turns out that my mom had read books at the same library. They might even have run into each other or pulled books at the same time from neighboring bookshelves. But the library was one thing. Citizenship was another. Did my mom want to be an American? She already had one child—me—who was born American, but my sister was not. She and my mom were both still Pakistani passport holders. How could she turn her back on that, she asked herself. What if something happened to her husband? Would she move away from the southern Colorado desert town we now called home? The Iranian woman at Apna Ghar was facing a similar but more pressing dilemma. Fortunately for her and for my mother, America stands for diversity as much as Islam does.

I don't know exactly when it happened, but I would like to think that one day my mother caught herself thinking of home and saw in her mind not the Pakistan of her childhood but our large, green backyard, my father's office just a few blocks away, my preschool building behind his office, the counter we ate breakfast at every morning. While Pakistan would always be the country of her birth, like the traders who traveled the Silk Road, the gallant Moors of Spain, and the early Muslims who fled to Abyssinia, she had made a new home too.

Four years after moving to Pueblo, she decided to become a citizen. My sister, Aliya, had received special permission to take the oath despite her young age. We were told that she was the youngest person in the state of Colorado to take the oath at that time. Their hands raised, they repeated after the judge, revoking any allegiances to any other country besides America. The judge gave a little speech welcoming the new citizens. "When I go out to eat dinner, I can have Chinese, Mexican, Italian, or whatever food I like. It's all because of immigrants like you," he gushed. My mom smiled a little bit wider that day. I knew she was proud to be an American.

No other country in the world—even Islamic ones—fulfills the diversity that is envisioned in Islam more than America. The gathering of Muslims at *hajj,* the annual pilgrimage of able Muslims, is a crowd with a variety of ethnicities. The crowd at the Colorado State Fair, held in my hometown every August, is just as diverse. The diversity heralded in Islam is achieved in America.

Soon after my sister's oath taking, we received a package. My parents were still asleep when it arrived, and my sister and I took advantage of their slumber to tear in to the box. It was addressed to her, after all. Inside was the most stunning thing I had ever seen. It was a bold satin piece of fabric. The red and white stripes gleamed out of the brown wrapping. I wanted to wear it like a dress, but my sister said it was not for wearing but for the Pledge of Allegiance. We draped it over a chair and, with our parents still sleeping, pledged allegiance to "one nation under God with liberty and justice for all."

Just before the Prophet Muhammad died, he gave his most important sermon to his *umma,* also on Mount Arafat, where the *hajj* pilgrims still gather today. He clarified many Islamic teachings in his speech, including Islam's prodiversity and equality stances. He reminded his fellow Muslims that we all came from Adam and Eve, that none of us is superior to the other: "[A]n Arab has no superiority over a non-Arab nor a non-Arab has any superiority over an Arab; also a white has no superiority over black nor a black has any superiority over white except by piety and good action."

He exhorted the community to see each other as family, as the *umma* he had devoted the last twenty-two years of his life unifying. "Learn that every Muslim is a brother to every Muslim and that the Muslims constitute one brotherhood," he said. "Nothing shall be legitimate to a Muslim which belongs to a fellow Muslim unless it was given freely and willingly," he continued. "Do not, therefore, do injustice to yourselves," Muhammad told the group, reminding them that they will have to account for their actions to God. He gave Islam its own Pledge of Allegiance, hours before his death: liberty and justice for all, under God and under Islam.

Chapter 6

A Woman's Religion

O mankind! Reverence your Guardian-Lord,
who created you from a single person,
created, of like nature, his mate, and from them
twain scattered (like seeds) countless men and women;
reverence God, through whom ye demand your mutual
(rights),
and (reverence) the wombs (That bore you):
for God ever watches over you.

Qur'an 4:1

"Does your father know you're doing this? What does *he* think?" I had just finished speaking about Islam in America at Colorado State University in Fort Collins, Colorado, and something I said had clearly upset this older Muslim man. In fact, I upset about half a dozen men that evening. Judging from appearances, they were Arab Americans, ranging from in their thirties to one in his fifties at least, and they had waited till the end of my talk to bombard me. I had spent the late afternoon driving north from Denver to the event. It was a crisp and sunny November day. The grass had started to turn brown at its roots, while the tips maintained a faint but robust shade of green.

Mainly these men were upset that I was not wearing the cover*, but they seemed to have other problems with me that I couldn't even begin to address. The wife of the organizer, a dean at the school, was accompanying me and began to intervene when I walked to within five inches of the gang leader and looked at him eye-to-eye, calm and collected. "My father does know what I am doing, and he supports it 100 percent. If you think I should wear *hijab* just because you tell me to, then you don't understand Islam at all," I said as authoritatively as I could. The event had generally been a success, and the expected protests were simply the cherry on the top. I know that Islam and the Qur'an support my independence as a woman.

Besides the fact that the Qur'an is against coercive proselytizing ("Let there be no compulsion in religion" (2:256), the Qur'an and Islam also firmly stand for women's rights. The Prophet Muhammad is personally responsible for the greatest advancement of women's rights in a single time period. He ended the practice of female infanticide. He also encouraged women to participate in politics and encouraged the tradi-

---

* "Cover" refers generally to a scarf some Muslim women choose to wear. A debate rages in the entire Islamic community whether the cover is required of Muslim women or not. Usually, the scarf or veil covers the woman's hair and neck but can also drape her chest and sometimes cover her face too. The cover is often made of black or white material but can actually be any color fabric. Although technically the accurate Arabic term for the cover is *jilbab,* terms like "the veil," *hijab, khimar,* and others are used to describe the cover.

tion of keeping maiden names after marriage. The Qur'an says that women should sign a prenuptial contract before they marry, which lays out their right to own property and to seek an education. These rights are God-given. The women in Muhammad's own life—his daughters and wives—were all strong, and he relied on them very much for decision making. In fact, he left no male heir but female ones, who assisted the community greatly in expounding his message after his death. The fact that individual Muslims have lost touch with the Prophet's example, like the gangs of Muslim men who often verbally attack me after my speaking events, does not make Islam against women. Every place where Islam has been practiced, the culture has affected how Islam is interpreted. Many of these cultures are patriarchal and violate the spirit of Islam.

In the winter of 1993, my family and I attended the wedding of my mother's first cousin. At all the South Asian weddings I had been to, the mother of the groom had a central role—usually by her own self-promotion but a central role nonetheless. She would present gifts to the bride in grandiose style; she would preside over all the ceremonies, especially the ones for her son, like a boisterous peacock. Everyone knew her son was the prize being offered and accepted. South Asian weddings include various ceremonies for the bride too. Other married women and especially the groom's mother ceremonially tend to the soon-to-be-married bride, rubbing oil or henna on her hands so her skin will feel soft with a soothing aroma for when she meets her husband and other future in-laws.

Naturally, I found it odd, then, when this mother of the groom had to be prodded by her sisters to join the bridal couple on the bridal stage on the day of the wedding for photographs. Parveen had been almost an absentee mother of the groom. Where other mothers had nearly surgically inserted themselves into the events of the wedding, my mother's aunt had blissfully been watching from afar. I wrongly assumed that she was acting out of maturity, that she didn't care about such mortal concerns as whether all the guests had admired her son's new suit sufficiently or pawed over the beautiful young bride marrying her son.

"Isn't it refreshing how *Khala* [Urdu and Arabic for aunt] takes a back seat? Usually the groom's mom is so in your face!" I said to my mother and sister after we had attended the wedding. We were now back at the home where we were staying.

"She's not taking a back seat!" my father interrupted from a connecting room. Surprised at my father's sudden interest in our girly gossip session, I looked at him perplexed. Had my father, who genuinely cares not for mortal concerns, seen something I had not?

"Prejudice won't let her participate!" he churned out with an impatient and irritated tone.

"What?" I said, confused, as my father marched off to his bedroom to change out of his suit. I looked to my mother for a translation (as I often did back then), and she said simply but not proudly, "*Khala* is a widow, and for her to participate is bad luck for the bride. The traditions are really meant for married women. Most people are too superstitious to let a widow be involved in their weddings. The symbolism is not good. Only a happily married woman should be near the bride."

"And *Khala* is too considerate—she wouldn't risk embarrassing the bride's family by participating," my sister, Aliya, added, clarifying yet one more aspect of South Asian culture that was bewildering for me but obviously crystal clear to her. When my mother's aunt sat on the stage with the bridal couple, she just sat off to the side on her own, nodding her head politely, agreeing with whatever someone on the other side of the stage was saying. This particular aunt was known in my family for having a master's degree—a big achievement in her time. She was and still is known for her strong opinions and political judgments.

As a young woman, she had picky tastes. Every morning, the family's chef ignored her complaints about her breakfast eggs: too salty, too sweet, underdone, overdone, too runny, not runny enough, and so on. The day of her wedding was no different. She immediately sent her first breakfast tray back to the kitchen of the Sahiwal, Pakistan, home she shared with her five sisters and one brother. This time, the chef humored her—not only because it was her day but also because she knew she must be nervous. On her own wedding day, about thirty years before her son's wedding where she sat so quietly, her complaints resulted in the house chef preparing her breakfast eggs roughly fifty different times, obliging her request for a new breakfast with each complaint. Eventually, the house kitchen ran out of eggs, clean dishes, and breakfast trays to accommodate her. They were all piled up in her bedroom where she reigned like a little queen, rejecting eggs almost as quickly as they were made. Wedding guests staying at the house ate the rejects, one by one, soon after word reached them of the bride's pickiness.

Some of the first converts to Islam were just like my great-aunt: strong, opinionated women who were widows. Arabian culture, in the Prophet Muhammad's time, like many cultures all over the world, treated women like property—valuable commodities when they could bear children and were untouched by a man, to be passed from father to husband until the husband's death when they became legal nonentities. As women, they could own no property and, without a spouse, were nothing but a drain, disdained by the pagan gods of pre-Islamic Arabia.

When Prophet Muhammad first began sharing his revelations with his community, many of these widowed women took notice. For the first time, they were being told that they had value. To the men of Arabia, the Qur'an said: "And if any of you die and leave wives behind, they bequeath thereby to their widows [the right to] one year's maintenance without their being obliged to leave [the dead husband's home]. If, however, they leave [of their own accord], there shall be no sin in whatever they may do with themselves in a lawful manner. And God is almighty, wise" (2:240). The Qur'an directed men to provide at least one year of support for their widows.

Furthermore, a woman was free to make her own choice to leave the home of her husband if she pleased, and the man would not be held responsible for her actions. The usual argument that men use—that their women's behavior reflects on their own honor and must thereby be regulated by them—was washed away by Qur'anic revelations.

Muhammad was the first leader in Arabian society to treat women as equals and not simply as property. His treatment of women was radical and revolutionary for its time—and still is. The rules my great-aunt had to observe were the vestiges of pre-Islamic, South Asian culture that still resonate today in Pakistan and resonate similarly in other Islamic countries. But Islam came to reform these practices and succeeded in many ways. For each time a Muslim man criticizes me out of his own pre-Islamic patriarchy, another Muslim man, maybe even two or three, responds like my father, scoffing impatiently at such traditionalism. With Muhammad as His mouthpiece, God said: "I shall not lose sight of the labor of any of you who labors [in My way], be it man or woman: each of you is an issue of the other" (3:195). Men like my father understand that the Qur'an treats men and women as equals.

Certainly other Qur'anic passages can appear to justify the contrary: that men are superior to women. Reading the Qur'an out of context and without an understanding of it in its entirety can result in negative views not just of how women should be treated but also of Islam generally. When I first read chapter four, verse eleven of the Qur'an for a college class on Islam, I thought it smacked of sexism: "Concerning [the inheritance of] your children, God enjoins [this] upon you: The male shall have the equal of two females' share"

(4:11). Wait! Was the Qur'an really saying that two daughters were worth one son? Was the antiwoman hype about Islam true? *How could my parents have not told me about this,* I angrily thought, *instead telling me only about all the good things Islam said about women?*

The next day in class I was ready for an argument, but I was not sure with whom. I was angry and felt betrayed. How was I supposed to have a fight with God? I tried to stay calm as my professor began her lecture, hoping that all the non-Muslim students wouldn't turn and stare at me. I was so upset that I felt like I was going to burn through my red-hot skin.

My college—being an all-women's college—was very feminist. My classmates would not tolerate this type of discrimination, and like so many times before in my life, I would be called upon to provide the perspective on what seemed inexcusable in Islam. Talking about the Crusades wars in Catholic school was one thing, but this inheritance issue directly affected me as a woman. The likelihood of my facing off with a Holy Roman Empire soldier was zero, but my ability to inherit was a viable and likely event that was now about to be put on trial by a jury of my peers—other college-age young women who would have sent the eggs back to the kitchen too.

I finally stopped panicking long enough to listen to my professor's lecture. She began by saying that certainly this passage of the Qur'an seems sexist. *No kidding!* I said to myself. Why were we wasting time? Let's just move on to the Inquisition and finish it off as soon as possible. What my professor said next took me completely by surprise. She claimed that the passage was not as bad as it looks.

She explained how another part of the Qur'an, which we would read later in the class, stipulates that a man must use his own wealth for all the household expenses—like maintenance of the home, cleaning supplies, clothing for the children and wife, food, and whatever else, including any other property he or his wife owns. A woman, on the other hand, is under no such obligation according to the Qur'an. Her money is hers alone in God's eyes. God will not punish her for not spending her money on food or clothing, even if her family clearly needs it.

The logic of the Qur'an then is that a man should inherit more than a woman because he will have to spend probably at least half his income on maintaining the household. An unmarried sister or widow could live in her brother's or father's house indefinitely under the Qur'an and without having to pay for any of the household expenses. God has already called for a portion of a husband's money. He must spend whatever is required to maintain the household whether he likes it or not and will be held accountable for not doing so. In yet another part, jurists have interpreted the Qur'an as saying that if a wife requests a housekeeper to clean the home and a nurse for their children, the husband must fulfill these requests at his own expense.

So, for instance, God would hold my future husband responsible for not supporting me financially, even if I told him I wanted to support myself. On the other hand, if I contribute or support myself, my actions are seen as charity by God and will be counted in my favor in the Judgment Day tally of good deeds—acts that I was not obligated by God to do but chose to do anyway.

As I see it, women have superior rights under Islam. Although they receive half the inheritance, they keep it all. They can spend it on the household if they like, and if they do, it will be a point in their favor on Judgment Day. But they can also spend it on clothes for themselves or give it away or keep it under their mattresses if they want, and it won't be held against them by God as it would for a man.

That religion class left me astonished and dumbstruck by my own religion. My slit-eyed anger had yielded to wide-eyed amazement. I hardly had time to take in the moment as my professor continued. The Qur'an and the Prophet Muhammad were among the earliest in recorded history to recognize a woman's right to own property. In Muhammad's time, women used to *be inherited,* not inherit themselves. Regardless of the portion, the simple fact that the Qur'an said women could inherit was a major revolution. "Men shall have a share in what parents and kinsfolk leave behind, and women shall have a share in what parents and kinsfolk leave behind, whether it be little or much—a share ordained—[by God]" (4:7).

Under Islamic law, women have had property rights since the seventh century. The Qur'an lays out specific portions to be given in a variety of instances—one daughter, two or more daughters only, widowed husbands, no heirs but with a brother or sister, and so on. "Thus is it ordained by God; and God is All-knowing, Most Forbearing" (4:11). I would have to go to law school and take Trusts and Estates years later to appreciate fully the comprehensiveness of the Qur'an. Even in the United States, with our various laws and cases on inheritance, we have not reached the simplicity and logic of inheritance laws in the Qur'an.

Muhammad, through the Qur'an, was one of the great lawgivers in the history of the world. As a religion student, I was naturally interested in law because most of our laws are derived from religious sources. The U.S. Supreme Court has a mural painting in its halls depicting these sources of law. Moses is holding the tablets of the Ten Commandments. Hammurabi is there with his code. Muhammad is there as well because, through his revelations, a way of life was given to humankind. Many of these guidelines advanced the position of women. Even by today's standards, the Qur'an is very progressive on women's issues.

Since college, three of my close friends have married. Two took their husbands' names. I see young women every day dealing with fallout from this decision. Once American women became free to keep their own names, the taking of their husbands' names became some kind of sacrificial rite loaded with meaning. Roseanne Barr took the name of one of her spouses because he converted to Judaism, her religion. Some women initially do not take their husbands' names but later do so after having a child who has the husband's last name. When I worked as a summer associate at my firm, I met another law student who had actually taken his wife's name. I've even heard of couples combining their last names to make an altogether new last name. Many women also hyphenate.

Fortunately, I will not have to wonder what to do when I marry. I am keeping my own name. This matter was settled a long time ago for me. In the seventh century, Muhammad upheld the custom that women should keep their maiden names, probably in recognition of the additional Islamic concept that women are not property but their own independent beings. I will also have a prenuptial contract when I marry, and not because I want to kill the romance or anything like that, but because my religion requires it. The prenuptial contract had its birth 1400 years ago with a revelation by the Prophet Muhammad.

The Qur'an was one of the first historical documents to recognize a woman's right to choose her spouse herself. It instructs Muslims to write a contract to govern their marriage—covering the dowry and whatever additional provisions the bride would like to add. Muhammad was also the first man to say publicly that a woman could choose her sexual partner and that she and her husband should only have sex with each other and in marriage.

Sexuality is a blessing in Islam to be enjoyed between spouses. Otherwise, sex outside of marriage, or with a woman one is not married to, could have negative fallout for the woman. She could become pregnant and face the harsh critique and mistreatment of her peers, while the father would be unknown and blameless. For that reason, any child born in an Islamic household is the responsibility of the male head of the household. No child would be fatherless in Islam.

Similarly, a woman is also free under the Qur'an to divorce her spouse: "If a wife fears cruelty or desertion on her husband's part, there is no blame on them if they arrange an amicable settlement between themselves; and such settlement is best, even though men's souls are swayed by greed" (4:128). Furthermore, in the event of a divorce, the husband cannot take back the dowry he gave his wife (4:20) or any gifts he gave her during the marriage (2:229). Divorce is described and stipulated for, even an arbitration-type proceeding is suggested if necessary (4:35). The Qur'an goes into even greater detail, specifying a three-month waiting period before the divorce can be finalized to make sure the wife is not pregnant (2:228), and it stipulates that it is the responsibility of the husband to provide for the wife through the pregnancy if she is (65:6) and to support the child according to what he can afford (65:7 and 2:233).

Islam may even permit abortion within the first trimester in the opinion of some Islamic scholars. A passage in the Qur'an describes the process of conception, noting that an angel visits the womb after the first ninety days and breathes life into what was before simply a soulless clot of blood. A life with a soul has begun at that point but not before. Some reports also say that women in Prophet Muhammad's time had abortions and that he did not object, which contributes to the view that abortion may be allowed. Certainly even a literal reading would leave the possibility open.

On the other hand, Islam is totally against the murdering of a child after she has been born. One of Muhammad's first acts as a prophet was to ban female infanticide. A common practice among pre-Islamic Arabs was the burying alive of baby girls. Like widows, they were seen as an economic drain and merely property. I am proud, as a Muslim, that the Qur'an addressed this horrifying practice head-on. The Qur'an strictly forbids killing children (6:151) and states that being ashamed of a daughter's birth and killing her is evil (16:57–59).

The Qur'an openly criticized the Arabs for engaging in such a hideous practice: "For, whenever any of them is given the glad tiding of [the birth of] a girl, his face darkens, and he is filled with suppressed anger, avoiding all people because of the [alleged] evil of the glad tiding which he has received, [and debating within himself:] Shall he keep this [child] despite the contempt [which he feels for it]—or shall he bury it in the dust? Oh, evil indeed is whatever they decide! [Thus it is that] the attribute of evil applies to all who do not believe in the life to come" (16:58–60). The merciful God of the Qur'an has little patience for those who engaged in infanticide: "Lost, indeed, are they who, in their weak-minded ignorance, slay their children and declare as forbidden that which God has provided for them as sustenance, falsely ascribing [such prohibitions] to God: they have gone astray and have not found the right path" (6:140). Those who kill their daughters will be accountable for their actions too (43:16–19). While we can disagree whether the other achievements related to women have fully bloomed, no one can contest that, in eradicating the practice of female infanticide, Muhammad was very successful. Islam is responsible for the massive and widespread end to female infanticide in Arabia and other areas where Islam spread.

The Qur'an's high esteem for women does not end with recognizing legal rights for them. The Qur'an instructs believers to respect, almost worship, their mothers. The Qur'an says to be kind to one's parents: "In pain did his mother bear him, and in pain did she give him birth" (46:15). The Qur'anic chapter devoted entirely to women says that all people should "reverence God . . . and the wombs (that bore you)" (4:1). Even though by non-Islamic cultural standards, my aunt Parveen is a nullity—she is, in reality, a matriarch of the family. Her widowed status, while having an effect on cultural events like weddings, has been elevated by Islam's emphasis on honoring mothers. Her own grown-up, thirty-something son bows to her authority. It's not even hard for me to see how someday the young and beautiful bride he married will exercise her own matriarchal power too. The natural cycle in many Islamic countries, despite the negative influence of pre-Islamic culture, is matriarchy: rule based on the mother.

I visited Pakistan again in the winter of 2000 for another family wedding—my own first cousin Nadia. At one of the family gatherings, the aunt who had picky tastes and a master's degree took me aside. We were in her home, where her now-married son and his young bride also lived with their two children. We sat on the sofa near the antique radio her husband would always listen to the evening news on. All I know of her husband is this radio—and also that he died from heart problems when his son was still very young and that he was quite stern.

When my mother was little and several of her aunts were still young unmarried girls, they would spend some afternoons visiting them in their summer home. One particularly mischievous aunt would drink from her brother-in-law's bottle of concentrated juice. The juice was a special mix of rose essence and herbs, specially designed to cool and soothe the effects of summer heat. But the brother-in-law, my uncle, was so strict that he would mark exactly the level of the juice so he'd know if some was missing. My mother was just a little girl and had no part in such deceit, but she would watch her aunt refill the bottle with water to the last marker line. Mention this story to my mother today, and she will imitate her long-gone uncle ("'Why does my *rooh afza* [rose essence] taste SO FLAT?'") and then giggle with delight at the memory.

"I am very proud of your book and what you are doing," my mother's aunt told me when she took me aside. "It's important to speak out and tell people about these things. Don't ever let anyone stop you," she continued. My great-aunt had no idea that one day I would be standing on four-inch stiletto heels so that I can look at a Fort Collins critic eye-to-eye. She probably also had no idea that, in all the glorious things that the Qur'an and the Prophet Muhammad have to say about women, the "cover" would be the hot issue it is among some groups. Even though she was part of and practiced a culture that treated widows as second-class citizens, she still is a strong woman. She is the way God meant her to be.

While the Qur'an gives Muslim women rights, it is the Muslim women themselves like my great-aunt who breathe life into the meaning of the Qur'an. What really makes women in Islam special are Muslim women, especially those who lived in Muhammad's time. I was in another religion class when I learned the greatest similarity between Muhammad and the Prophet Jesus, whose teachings Muslims also believe in. As a religion major, I was required to take a Bible-study class, and I chose one on the New Testament. Struggling to stay awake in the 8:30 a.m. class, something the professor said caught my attention. "What we know about Jesus historically, we know from the women in his life," he said.

The historical accounts of Jesus by his contemporaries had been recorded and then passed down by women. I suddenly realized that the same was true of Muhammad. Anyone who wants to know about Muhammad must learn about the women in Muhammad's life. Muhammad led a community filled with active and strong women. In fact, when you read about the women Muhammad knew and interacted with, especially his wives, you realize that if the Qur'an did not grant women all the rights it did, these women would certainly have taken these rights for themselves anyway!

As I pointed out earlier, Muhammad, the messenger of Islam, married his boss. His first wife, Khadijah, was his employer. Although he was a businessman in his own right, the older woman proposed to him. He accepted despite the age difference. He worked for Khadijah's very successful caravan trading company. She was an essential partner in Muhammad's religious life too. After receiving his first revelation, Muhammad ran home to her, in a panic. The angel Gabriel had come to him, ordering him to "Read." God was trying to speak through Muhammad, but Muhammad needed to give in and let God's words flow out of him. Scared that he was losing his mind, Muhammad ran away to the safety of Khadijah and asked her to cover him. She held him and asked him what had happened. She recognized the name of Gabriel and immediately took the *shahada* (profession of faith) that all Muslims believe and that a non-Muslim must say to convert officially: "There is only one God, and Muhammad is his messenger."

It was Khadijah who recognized Muhammad's prophecy first. Without Khadijah, a woman, Islam would not exist. One of the world's greatest religions was built upon the belief of a woman in her husband, when even he didn't quite believe in himself.

Muhammad was married to Khadijah for twenty-five years—the entire time he lived in Mecca before the exodus of the small Muslim community to Medina in the *hijra* in 622. The year before the *hijra* was a difficult one for Muhammad mainly because of Khadijah's death. This dynamic woman, who had been his closest companion and ally in the struggles of early Islam, had died. Fortunately for Islam, he carried on, but the example of Khadijah is a strong one that is an inspiration for young Muslim women today.

Islamic scholar Karen Armstrong writes that Muhammad greatly loved and admired women. Another Islamic scholar, Fatima Mernissi, reports that Muhammad publicly stated that he preferred the company of women to men. According to Mernissi, Muhammad was attracted to them and wasn't ashamed of his attraction. His wives and female friends were among his closest advisors. One of his wives always accompanied him on any military expeditions he took. After Khadijah's death, he married several women (although he took no other wives when she was alive). Although most of these marriages were to form political alliances with other tribes or communities, he enjoyed a sexual and affectionate relationship with nine of his wives and had many children. (All of Muhammad's children died young except for his daughter Fatima, who became a leader in early Islamic history.) He also married several widows so that they and their children would be provided for.

His actions sent the message to the community that supporting one worse off than oneself was an admirable and worthy goal and that no stigma existed in marrying a widow. He never let his own position as leader of the community cause him to lord over his wives either. He is known to have helped with the chores and mended his own clothes. Even his own male companions were surprised to find out these facts, and we probably know them now because women who knew Muhammad passed them down.

Muslim women have been religious teachers since the beginning of Islam, and many were from the Prophet's own family. Muhammad's wife Aisha is said to have transmitted more than 2,000 *hadiths,* although only 300 have been authenticated. Aisha may have been as young as seventeen when she married Muhammad. Her young age allowed her to live for forty-six years after Muhammad's death and, in that time, educate the young Islamic community on the true meaning of many Qur'anic passages and on the life model the Prophet set for them. Aisha also settled many disputes over *hadiths* of the Prophet with her own firsthand knowledge of the Prophet's habits and behavior. She led one side in Islam's first civil war. I would say that Aisha—a woman—was the first scholar of Islam. Muhammad had no male heirs but several daughters, including Fatima, who played a role similar to Aisha's in early Islam.

Muhammad encouraged the women of the early Muslim community to participate in local affairs, and under his watch they served in combat. These women asked Muhammad questions about the Qur'an that resulted in direct revelations. Perturbed that the Qur'an had so many male pronouns but none for women, as if the only believers were men, the women of early Islam asked Prophet Muhammad about this. Soon after,

Muhammad received a revelation that placed women on an equal footing with men. The verse contains eleven lines that say that both men and women who believe in God and are good will be favored by God: "for [all of] them has God readied forgiveness of sins and a mighty reward" (33:35). From that revelation on, women are explicitly addressed in the Qur'an.

Like with any religion or culture, the advances made by women must be vigorously maintained and fought for. You don't need to read the paper today to know that Muslim women are mistreated. As a young adult growing up in the 1980s, I think the question I heard most from people who had just met me was: "What is it like to be a Muslim woman?" They asked as if I had struggled for my freedom, come to America on a ship with half-torn and tattered clothing, yearning to breathe free air. Aside from the fact that I was born and raised in America, a fact people still seem surprised at, I never struggled for freedom because both my country and my religion already gave it to me—both as a woman and as a Muslim.

The stories we hear in the news, things we read on the Internet, do come from somewhere, but not from Islam. Just because a Muslim does or says something doesn't mean it is automatically Islamic, especially when it comes to Muslim women. A perfect example is the Qur'anic injunction to men that they should leave women alone when they are menstruating—mainly in the sense that they should not have sex with them (2:222). Over the years, patriarchal Muslim men looking to increase their importance have used this Qur'anic passage to justify all kinds of exclusion and discrimination based on women's menstrual cycles. No evidence exists that Muhammad similarly excluded women who were menstruating, but many men blindly point to this passage as justification for all kinds of discrimination.

The Qur'anic provision for polygamy has also been manipulated by some Muslim men. Many Muslim men engage in polygamy that is illegal in the country in which they reside, which defies the Qur'an's mandate to abide by the laws of your country. In addition, scholars of the Qur'an agree that polygamy was a temporary Qur'anic solution to the problem of widows and orphans. In order that a woman not be homeless or poor, Islam allowed a man to marry a widow, particularly if she had children who were without protection and stability. Polygamy was a conditional allowance, a measure to be used only to prevent the infliction of injustice on orphans.

More important though, the Qur'anic passages on orphans (4:3 and 4:129 among others) clarify that a man can only have up to four wives and only if he can treat them all equally. The astute Muslim man would also note that the Qur'an says explicitly that a man is incapable of treating more than one wife equally. In the end, the view that a reasonable person would take away is that polygamy is no longer allowed. However, reports of Muslim men in the United States and abroad having more than one wife abound. Even as to the simple number, these men have violated the Qur'an already, with some men in Islamic countries having wives in the double digits! As with the passage on menstruation, the regulated and once-needed use of polygamy endorsed by the Qur'an has been expanded upon by some Muslim men impermissibly.

A few years ago, several of my "friends" forwarded me articles about a Nigerian woman who was going to be stoned as punishment for adultery, per the Islamic law this court claimed to follow. (The woman, Amina Lawal, has since been found not guilty by a higher court and spared the death sentence. However, other women have suffered and continue to suffer under the tyranny of so-called Islamic courts claiming to expound Qur'anic law.) These "friends" outwardly asked for my thoughts, but it seemed like they were really trying to shake my faith—a variation on the "What is it like to be a Muslim woman?" question.

I felt terrible for this Nigerian woman. She probably hadn't even done anything wrong. But I never doubted that what was happening to her had nothing to do with Islam. I knew, from my first reading of the article, that this rural court was not practicing Islam, but people reading the articles circulating on the Internet would not know that. If the court were truly following the Qur'an, it would not be ordering stoning. Although it sounds harsh, the actual punishment for adultery in the Qur'an is flogging, not stoning (24:2). So if this court were so representative of Islam, it would not be ordering stoning. Beyond that superficial point, though, in order to prove adultery in a court governed by the Qur'an, the accuser must produce four witnesses to the actual, sexual penetration (4:15). Speculation and hearsay do not suffice.

As a result, and many scholars have pointed this out, no one would ever actually be punished under this Qur'anic provision because the likelihood of producing four actual witnesses to the penetration is so slim. So, as harsh as the flogging punishment sounds, in reality it was probably never meant to be enforced. The flogging is a legal fiction of sorts, designed to discourage adultery but unenforceable on its face. The Qur'an goes further, adding that if the accuser cannot produce four witnesses, or if the witnesses are found to be lying, the false accuser will be flogged instead (24:4–5). The goal, then, is that no one make frivolous accusations. The Nigerian rural court was acting on its own but using Islam as general justification. Because few people know what the Qur'an actually says, the small-minded or ill-intentioned can easily speak about Islam with unchallenged authority. The same forces that prevent my great-aunt from fully participating in a wedding function continue to keep women oppressed in the name of Islam.

The horrifying practice of female genital mutilation (FGM) is an even better example of how local culture gives Islam a bad name. FGM is the cutting of a woman's clitoris or other genitals to decrease her sexual pleasure. FGM is also called "female circumcision." Though some Muslims have erroneously engaged in FGM for years, it is nowhere near an Islamic practice. FGM is not mentioned anywhere in the Qur'an and would probably be best analogized to female infanticide—which is explicitly banned by the Qur'an. However, many Muslims and several non-Muslim communities continue to engage in the practice despite obvious problems with it, like infection and massive pain to the girl receiving the surgery, not to mention the long-term scars on her sexuality.

Simply, FGM had been practiced since long before Islam existed. But once the communities that do FGM became Muslim, they continued it, wrapping it up in general traditional behavior. To this day, FGM proponents will only say that FGM is their tradition, but if pressed, they cannot find an Islamic justification for it because none exists.

Muhammad meant to eradicate these practices, and he succeeded—mostly. However, all of us—Muslim or not—still struggle with patriarchy. Culture is a very powerful force that cannot be easily changed. The job of Islam and Muslims to reform these practices is not done, but not for lack of potential or desire on Islam's part.

With such a rich tradition of strong Muslim women and such clear language in the Qur'an emphasizing women's rights, I know where I stand as a Muslim woman. I don't need to defend my position or constantly fight for it. Like my great-aunt, I can just be the woman God meant for me to be: free to have an education, to own property, to be judged as a man would be, to be equal in God's eyes, to marry, to divorce, to vote, even simply to keep my maiden name.

Out of scores of positive, even glowing references to women in the Qur'an, maybe a handful could be interpreted in a way that demeans women. For instance, one passage reportedly instructs husbands to beat lightly wives who continue to be disobedient. As shocking as this passage is to me, I was relieved to find alternate translations by prominent American Muslim scholars—one an African American Muslim and the other of South Asian descent like me, Amina Wadud and Rifaat Hassan, respectively. These alternate translations show that the word for "beat" in Arabic ("baraba") has several different meanings. In fact, of the many times this particular Arabic word is used in the Qur'an, in only one instance does it indisputably mean to hit. In fact, the most used meaning is "to walk away." Other meanings also used are "to neglect," "discard," or even "explain." This word probably did not mean "beat" in Muhammad's time, when he first received this passage from the Qur'an.

These alternative translations fit in with the Qur'anic view of women more accurately, and American Muslim women who are scholars are the ones uncovering them. In reevaluating and challenging what has been accepted about how Islam treats women, we are actually making Islam stronger.

One of the most famous Muslims of all time is Rabia, a Sufi poetess who lived in the eighth century. She was born in Basra and was not concerned with the issue of women's rights under Islam. Her contemporaries still admired her for her unprecedented faith and deep mysticism. Like the Prophet Muhammad, she had also been an orphan. Her parents had died when she was young. Although she is reported to have come from a wealthy family, the circumstances of her parents' death left her penniless. Sold into slavery, she had nothing, except for her faith in and love for God. According to legend, her owner watched her praying one night and saw a halo appear around her head. Not wanting to catch trouble with God, the first thing he did the next morning was free her.

It was her piety, her belief in Islam, that brought about her liberation. She then lived an ascetic lifestyle, making it evident that all she needed was God. A famous story about her is that she was seen walking alone and quietly with fire in one hand and a bucket of water in the other. When someone asked her what she was doing, she said, "I want to throw fire into Heaven and water into Hell so that both will disappear, and we can contemplate God alone."

In the Qur'anic story of Creation, Eve is described as being created independently of Adam and not from his rib (39:6), and they both coequally participate in being tempted by the serpent to eat of the forbidden tree. Like Rabia, Eve, the first woman, walked on her own. In Islam, men and women, like Adam and Eve, share everything equally. As the Prophet Muhammad said to the menfolk in his last sermon, just before his death at a time when he could have spoken about anything else, "O People, it is true that you have certain rights with regard to your women, but they also have rights over you."

The Qur'an says that men and women were created to be a comfort to one another—not an additional source of division (30:21). Men and women are equal in the struggle to be good: "Whoever works righteousness, man, or woman, and has faith, verily, to him will We give a new life, a life that is good and pure, and We will bestow on such their reward according to the best of their actions" (16:97).

When I had finished with the Arab men in Fort Collins who wanted me to wear *hijab,* the organizer and his wife accompanied me to my car. Through the night I drove back to my home, and in the dark I could no longer see the brown-green grass of Colorado but only the dark blue night sky filled with stars. Like Rabia, I wished then and still sometimes wish that we could extinguish the fire of Hell and the promise of Heaven—the arguments over *hijab,* the fiendish interest in the oppression of the Muslim woman. Then maybe we would fully realize what God means for us all, without the drama and minus the hype. Perhaps like Rabia, in the dark night, with only the road ahead, we would only see God.

promise of Heaven—the arguments over which [illegible] [illegible] [illegible] in the [illegible] of the [illegible] [illegible] [illegible] [illegible] would [illegible] realize what God means [illegible] [illegible] the [illegible] [illegible] name the type. Perhaps like [illegible] [illegible] the dark [illegible] with [illegible] the road ahead he would only [illegible].

Chapter 7

Being Muslim Makes Me a Better American (and Being American Makes Me a Better Muslim)

It may be that God will grant love (and friendship)
between you and those whom ye (now) hold as enemies.
For God has power (over all things);
And God is Oft-Forgiving, Most Merciful.

Qur'an 60:7

But it is possible that ye dislike a thing which is good for you,
and that ye love a thing which is bad for you.
But God knoweth, and ye know not.

Qur'an 2:216

"You guys were supposed to come in the back door, *not this door!*" I looked up at my mom from the entryway buffet, where I was busy eating cheese cubes and trying carefully not to drop anything on my *gharara* (a Pakistani ethnic gown, consisting of an embroidered tunic top, scarf, and very wide-legged pants, mainly used on dressy occasions and weddings).

"Mom," I said in response, "all hell is breaking loose out here!"

"Well, all hell is breaking loose in there," she said, pointing to the crowded banquet hall. My sister, my poor, defenseless older sister, was trapped under a canopy being carried by six "happily married ladies." It was her wedding day, and as with all weddings, even the best-laid plans manage to go awry.

To the outside observer, the wedding was a beautiful event—a re-creation of Mughal splendor. My sister's gown was a true cross-cultural creation: French bridal lace purchased in Denver, Colorado, dyed a peach-salmon color specially created for the dress by an Indian artisan whose technique dates from the Mughal period in water from the Ganges river that Hindus worship, embroidered in Delhi with small silver European crystals by octogenarian seamstresses who had been sewing since before they could walk, and designed by an up-and-coming Asian designer from her home in Connecticut. And now it was being worn at a wedding in Las Vegas. The scarf was pinned to my sister's hair, and the skirt of the gown just touched the floor. It was perfect. Even Elvis had never looked this great.

Even though my mother had shown us during rehearsal the night before the door we would enter from, which conveniently adjoined our dressing room, in all the excitement the next day I totally forgot. In the chaos, the various staffers directed us to the entrance the guests had used to enter the banquet hall. The entire bridal procession, which included me holding the Qur'an, my brother, my sister's bridesmaids, and, of course, the "happily married ladies" accompanying the bride, and the bride herself, was stuck for several minutes in a fast-forward-then-rewind, back and forth motion depending on which staffer was currently prevailing in the argument. ("That way!" "No, that way!" With each exclamation, we'd move in that direction and then back again with the next.) When we finally reached the entrance to the banquet hall, I set eyes on the hors d'oeuvres. I had become famished in the effort and set about for a snack.

"Asma, what are you doing?" my sister, the bride, scoffed.

"I'm starving!" I whined. Then my mom discovered us. We processed in soon after that. I reached the front of the room where the groom was waiting, handed the Qur'an to my grandmother who had been waiting to receive the bridal couple, and then sat off to the side where chairs had been left for us. The bride soon entered with the happily married ladies, whom I had handpicked.

Wherever Islam is practiced, including in America, the Muslims there have used or added their own culture to its practices. Some of these practices are harmless, like the happily married ladies of South Asian weddings, while others can cause non-Muslims to confuse non-Islamic, cultural practices (such as FGM, polygamy, and so on) with Islam. Among the great Islamic empires, very few were able to minimize the effect and influence of local culture. While Islam was the religion, other matters—including culture, law, and politics—were influenced as much or more by non-Islamic culture that had existed before Islam. Today, a wedding in Saudi Arabia will differ wildly from a wedding in Malaysia even though both are technically Muslim weddings. This cultural diversity in Islam can be a difficulty but can also be a blessing.

Although Muslims all over the world vary widely in interpretation because of this cultural drift, Islam is a richer and deeper religion for it. The cultural variety among Muslims is to the credit of Islam too, showing how easily adaptable the Islamic faith is. The fringe benefit of practicing Islam with a touch of American culture is that the aspects of Islam often overlooked in Islamic countries—like racial and gender equality, racial harmony, charity, and humanitarianism—are rediscovered by American Muslims. These aspects are a part of American culture, and the dual emphasis from Islam the religion and America the culture allows American Muslims to practice a more Qur'an-based Islam than many world Muslims.

For instance, I have attended many South Asian weddings. As tradition dictates, the bride processes in accompanied by women who are "happily married." When I was a teenager, after attending such a wedding, I asked my mom why the brides weren't given away by their fathers like in Christian weddings on television. "The happily married ladies are like a good omen," my mom explained. Their presence around the bride is meant to bless her wedding and marriage so that she will have a happy marriage like they do. The happily-married-ladies tradition most likely is borrowed from South Asian Hindu culture.

"But Mom," I said, "I know the women who were in the wedding, and I don't think any of them has a happy marriage!"

"Well, just don't say anything, for goodness sake!" my mother whispered, although we weren't at the wedding anymore. So, I vowed, years later when my sister began planning her wedding, that her happily married ladies would actually be happily married. We had my mother's aunt Babo, who has been married quite happily to her husband, Javi, since before Aliya's birth. We had Babo's daughter Tina, who has been similarly happy. We had an old family friend named Shireen, who is so happily married that she has a small gang of children, and so on. Babo and Shireen were the tallest, so they held up the rear of the canopy. The bride is covered with a canopy mainly out of tradition. At one time, these weddings were held outdoors, and to protect her from the weather, the bride was never left uncovered. Now the canopy has become a part of the ceremony.

While the happily married ladies were quite fun and added to the festivities, the requirement that they be present was a cultural one, *not* a religious one. Nowhere in the Qur'an does it say that happily married women must accompany a Muslim bride when she walks down the aisle. It's a nice touch, but it serves no theological purpose. But anybody unschooled in Islam would probably assume that all these cultural aspects were actually required by Islam. In fact, some Muslims probably think these traditions are Islamic. But they are not.

In terms of clothing, food, tempo, and design, a Pakistani Muslim wedding has more in common with an Indian Hindu wedding than it does with any other kind of wedding, including Islamic ones. Of course, the two weddings aren't identical. The religious ceremonies in the weddings differ greatly. The Islamic *nikah* ceremony is the same for all Muslims regardless of where they are located and has no similarity to the religious portion of a Hindu wedding (or one of any other religion). But the fun parts of a wedding—the colorful, cultural aspects—differ depending on local culture. As a result, South Asian weddings—whether the participants are Muslim, Hindu, Sikh, or Christian—look and feel a lot alike!

Out of all the cultures in the world, however, true Islamic values, as embodied in the Qur'an and the life of the Prophet Muhammad, most closely resemble American values. Although it may surprise many to read this, Islamic and American values are similar and enhance each other. For instance, the requirement to do charity—called in Islam *zakat*—is an American value as well as an Islamic one. The Qur'an says of charity: "For those who give in charity, men and women ... it shall be increased manifold (to their credit), and they shall have (besides) a liberal reward" (57:18).* Charity is, in fact, one of the five pillars of Islam; all Muslims are required to donate a portion of their income to the needy. In America, we are also encouraged to help those worse off than we are.

Today, when I practice *zakat,* I do it as both a Muslim and an American. As a Muslim, I am required to donate my time or money to charity. As an American, I am taught that I should help my fellow woman and man in need. Both Islam and American culture encourage people to be active members of society. Both Islam and American culture teach us that we owe it to our communities to be involved, not to sit back and let changes happen as they may. Muslims and Americans both believe in working hard and contributing to one's com-

* Other passages on charity include Qur'an 9:71 and 23:4 among many others.

munity. Muslims in America are freed from the same culture that once oppressed Muhammad and the first Muslims, living in a society based on and guided by the same themes that Islam was founded for. Ultimately, being a good American is the same as being a good Muslim. Islam stands for a just society, where people are valued and treated well—the kind of society we create and strive for in America.

In fact, Islam in America is, for the first time in a long time, an Islam freed from the pre-Islamic Arab culture that is largely un-Islamic and that inhibits the true goals of Islam. For every few benign traditions like the happily married ladies, the culture of Muslims includes one vestige of pre-Islamic, tribal culture that holds those Muslims back. These negative traditions not only prevent some Islamic countries from fully developing, but also prevent those Muslims from truly practicing Islam. In America, Islam has a chance to fulfill the values it really stands for, without politics or patriarchy to hold it back. Nearly every American Muslim I know feels that America is the only true Islamic country—that stands for the values Islam does—a fair and just society like the one Muhammad created in Medina. When I exercise my right to vote, when I meet with people of a different racial descent, I am confirming both what Islam stands for—women's rights, political participation, and racial equality—and fulfilling American values as well.

When I finally processed into the banquet hall, I could see why my mom had said that all hell was breaking loose. No one was sitting in his or her assigned seat. The seating chart I had worked on so diligently until that afternoon was being openly defied! Had I not been exhausted (and still famished), I would have been livid.

The guests at my sister's wedding had given in to the temptation to change their seats around. The story I heard later was that one of my sister's male friends from medical school liked one of her female friends from high school. So he exchanged his name card with the one for the spot next to her. When the person who had been moved discovered that his seat had been changed, he decided to ignore the seating altogether. From there, the seating arrangements, especially among the younger guests, spiraled out of control. Their naughtiness is reflective of how Satan is described in the Qur'an. Karen Armstrong describes him as mischievous but manageable, a clear contrast to the Catholic understanding of Satan, which is far more damning.

Armstrong argues that the Qur'an says God will forgive Satan on Judgment Day for his failed attempts to lure humans into his mischief. So when the Ayatollah Khomeini called America "The Great Satan," Armstrong feels, the Ayatollah was comparing America to the naughty and irresponsible figure in the Qur'an, and not the evil, dark Satan of the Bible. Certainly, even being called the Islamic version of Satan is not a compliment. But I would rather be compared to the guests at my sister's wedding, who made a mockery of my beloved seating chart, than to the condemned Satan of Catholicism.

✶

"Do you think you will move back when you are older?" people often ask me. Move back where? I want to ask them. First of all, I have only known America. I have visited Pakistan as much as I have visited Europe. On that basis, perhaps I should move to Europe. Second, my parents don't just come from Pakistan. They come from all over. My father's family left India at the time of the partition and came to Pakistan—so does "back" mean Pakistan or India?

My mom's family has wandered throughout Central and South Asia for years, leaving branches like Hansel and Gretel's crumbs in Uzbekistan, China, Mongolia, Afghanistan, Iran, and Pakistan. Which of these countries would I move back to? Because my parents emigrated from Pakistan, some people assume that I have a loyalty or allegiance to Pakistan. I suppose that I am more interested in Pakistani news than, say, news about Ireland, and I would probably root for the Pakistan team in the Olympics over a country I don't know at all. But I certainly have no plans to move there. If I could no longer live in America, I would probably move to Mexico—because of its proximity, its culture, and because I speak Spanish. I appreciate Pakistani culture and grew up around it, but I am more a product of American culture than any other.

I was the first child in my family to be born in America. Non-Muslim Americans often emphasize this point when they are introducing me before a speaking event or describing me to someone: "She was born in Chicago and raised in Colorado." At first, I thought the particularity with which people would point out my American birth was sort of quaint or interesting. But after hearing the pointing out so many times, I now want to say, "Get over it!"

Is it really that odd that a young Muslim woman who practices her religion was born in America? Is our opinion of Islam so low that we can only imagine those unfortunate enough not to have been born in America as willing to practice it? I have been American my whole life and Muslim my whole life, and I've never had a problem with being either. Others—non-Muslims—have had the problem.

Like all American schoolchildren, I learned about Columbus's journey to America. He was looking for India, though, not America. When the Native Americans greeted him on the shore, he called them "Indians" because he thought he had found India. The purpose of his search for India was to overcome Muslim control of the East Indies spice trade. The "discovery" of America was motivated by a desire to outflank Muslims. How funny that in pursuit of Spain's policy against Muslims, Columbus, who had been commissioned by Spain, inadvertently found America? Years later, America would become home to over seven million Muslims, maybe even as many as twenty million.* The country that Columbus has been credited for "discovering" would stand for the same principles of the religion his voyage was meant to undermine.

---

* We do not know an exact number because the U.S. Census does not ask about religious affiliation, but a conservative estimate puts Muslims in America at seven million. That estimate is based on immigration from Islamic countries and mosque attendance among other factors. The problem with such an estimate, though, is that it may not include new immigrant populations and periphery communities like Bosnian and Albanian Muslims and Muslims who do not regularly attend a mosque.

What surprises many Muslims and non-Muslims alike are the many striking parallels between the principles of Islam and the founding ideals of the United States. Just as America has the Constitution, Islam has a Constitution as well. Muhammad wrote a constitution for Medina in 622, spelling out the rights each citizen can expect. The Constitution of Medina specifically stated that the community consisted of Muslims, Jews, and pagans. (No Christians lived in Medina. If they had, the Constitution surely would have included them too.) The Medina Constitution required that all the groups had to look out for one another and were now, under the Constitution, bound together as citizens of Medina. This religious diversity was written into the founding document of the first Islamic community. Muhammad himself wrote it—a sort of premodern "We the People." The American Constitution was written in an effort to unify a band of people who had fought their own oppressors and were now trying to make a new start. The motivation behind "We the People" exists in both Islam and America.

The Prophet Muhammad, in his last sermon before his death, emphasized the inherent equality of all people: "All mankind is from Adam and Eve, an Arab has no superiority over a non-Arab nor a non-Arab has any superiority over an Arab; also a white has no superiority over black nor a black has any superiority over white except by piety and good action … Do not therefore do injustice to yourselves." He closed by telling the Muslims that they were all brothers to one another and that all were equal: "You are all equal.

Nobody has superiority over the other except by piety and good action."

In this final sermon of Muhammad's, the most important values of Islam were laid out, including the ones above—property and racial equality—as well as others such as women's rights. These values are the core values of America, too, where the founding fathers wrote a Constitution (the ideals of which took time to manifest fully) that said all people were entitled to own property freely, to exercise their political rights, to vote—simply to be free people: "We the people of the United States, in order to form a more perfect union, establish justice, insure domestic tranquility, provide for the common defense, promote the general welfare, and secure the blessings of liberty to ourselves and our posterity, do ordain and establish this Constitution for the United States of America."

From this example of religious freedom and racial equality, Muslim rulers, from the Prophet Muhammad onward, protected religious minorities who lived in their empires. They did not ask them to convert either. Given the title of *dhimmi,* these religious minorities, in exchange for a tax assessed on them, enjoyed the protection of the Islamic armies. If the minority community alone was attacked, the Muslim armies would fight on its behalf, even if not one Muslim had been attacked. The belief that each person has a right to practice his or her religion is an Islamic and American one.

The *dhimmi* classification has been cited as proof of an alleged Islamic hostility toward minorities. In fact, as noted in chapter 5, the *dhimmi* status was revolutionary for its time. Until then, no empire had treated minorities as anything other than political pawns, to be oppressed, abused, or celebrated as was politically expedient at the time. Of course, certain Muslim rulers may have lost touch with the proper use of this concept, but in its intent the *dhimmi* categorization was far ahead of its time. *Dhimmi* saved the lives of many Christians and Jews in history, whether anyone wants to admit it or not. For instance, Christians and Jews prospered in Islamic Spain at a time when they were persecuted in other parts of the world. Islamic Spain of the tenth century onward was a multifaith empire that counts the great Jewish philosopher Maimonides among its citizens. In addition, some Roman Christians in the Byzantine Empire in the fifteenth century and earlier, who had been persecuted by Greek Orthodox, welcomed and enjoyed Muslim rule. The Ottoman Empire of the sixteenth century was the most tolerant European regime of its time.

The oppression of religious minorities by Islamic regimes today and in the past is against Islam, not part of it. Of course, simply because Muslims are behind it does not make such tyranny Islamic. Religious discrimination by a Muslim is even less Islamic than the "happily married ladies." The history of Islam and the example of the Prophet stand for a recognition and protection of minorities. This principle is more in line with the American spirit against discrimination and for religious freedom than with the policy of many Islamic countries.

Like George Washington, Muhammad could have easily established a monarchy in Islam. But Washington and

Muhammad were revolutionaries. Washington didn't want to be king, but a leader chosen from among the people, a democratically elected leader. Centuries before Washington was even born, another man had also proposed that the leader of his community be chosen by the community, selected in a democratic fashion. At his death in 632, he instructed the Muslims of his community to choose a leader from among themselves*—probably the first time in Arabia and many parts of the world that a community leader espoused the ideal of democracy. Muhammad left behind the Qur'an, a holy book that set out a way of life with respect for all, for all time, and to be reinterpreted as times changed and as the people who followed it spread out all over the world. The Qur'an called for these reforms, which is the word of God to all Muslims, and the man who called for democracy 1,400 years ago was the Prophet Muhammad.

Both Americans and Muslims face the challenge of living up to the promises of their belief systems.

✶

* Muhammad specifically did not appoint a family member to follow him. Later, after his death, a portion of the Muslim community supported the leadership of his cousin and close companion, Ali. Ali also was the Prophet's son-in-law. They felt that a relative of the Prophet's *should* lead the community. The disagreement caused this community to form their own political party: the Shi'ites. Iran and Iraq today are both majority Shi'ite populations, although Shi'ites represent only about 10 percent of the world Muslim population and the American Muslim population.

Some scholars say that Thomas Jefferson and George Washington, among other founding fathers, had copies of the Qur'an in their libraries—the ideas of democracy and equality presented in the Qur'an are so strikingly similar to the foundational ideas in the U.S. Constitution. In fact, Muslim congressman Keith Ellison was sworn into office using a Qur'an once owned by Jefferson, currently kept in the Library of Congress.

The pagan society of Mecca before and when Islam first arrived was a cruel one. Only those with money were treated well, and those who had money abused everyone else. Among the Qur'an's first commands to the early Muslims and Muhammad was to establish a society based on justice. People were commanded to treat each other better than they had been. Widows and orphans who had been mistreated and left homeless, women who had been considered animals or property, the entire community that had been focused on drawing in more money and trade, with no one caring for those worse off—Islam commanded that this view change for good. The Qur'an was a declaration of independence in itself—the word of God manifested on Earth, saying that all humankind is created equal.

Ramadan is a striking example of this commitment to social justice. One of its purposes is to help Muslims identify with those who go hungry every day, to have them suffer as the hungry suffer. Simply giving to charity is not enough. We must starve ourselves. A socially just society where everyone is treated equally was the goal of the Qur'an and the early

Muslims and is still the goal of Muslims today. Every year, during *Ramadan,* we do not drink or eat anything from just before sunrise to sunset. We don't necessarily enjoy fasting, but we are required to do it.

Ramadan is meant to influence Muslims in the Qur'anic teaching that focusing on one's own fortune is wrong. The Prophet himself fasted once a week—a practice that medical science now says can lengthen one's life and improve one's health. The certainty of knowing that you will not be able to eat is a humbling one. God wants Muslims to be humbled like this—not just to appreciate what we have, but also to share what we have with those who have less.

The Prophet Muhammad was one of those people who had less. His own story reads like the American Dream. He came from nowhere to become a successful businessman. He was an orphan and poor. Though the tribe he belonged to ruled Mecca, he belonged to the lowest subtribe. His mother died soon after his birth, and he would have died as a defenseless infant had his grandfather not chosen to adopt him. He was a small-sized boy and, reportedly, was picked on often and selected to do the most menial tasks. He never had enough money to go to school and was illiterate his whole life.

By the time he received the first revelation of the Qur'an in 610, he had overcome the problems he was born into. He was a successful businessman who married his much older female boss. He was known for his honesty and integrity and had a reputation all over Arabia as a good negotiator.

Muhammad was a capitalist who made his living off of caravan trading. As in the United States, one of Islam's great reforms was the recognition of individual property rights, where each person kept what they owned. Just hours before the Prophet Muhammad died, he reminded the early Muslims of some key teachings in what is called his last sermon. He delivered this important sermon (*khutba* in Arabic) of Islam on the same Mount Arafat near Mecca where Hajira ran between Safa and Marwa, where Adam and Eve first found each other, and where millions of Muslims visit each year. He told the early Muslims to safeguard property rights: "[R]egard the life and property of every Muslim as a sacred trust. Return the goods entrusted to you to their rightful owners. Hurt no one so that no one may hurt you."

Muhammad also wrote a constitution for the city of Medina, recognizing democracy and rights for minorities, including Jews. By constituting social reform, a man who was once illiterate and an orphan became one of the greatest leaders in history. Before he even founded Islam, he had already turned his life around. He was set, by the circumstances of his birth, to fail, but instead he became a success at the caravan trading business. Becoming the founder of one of the world's greatest religions with a population of over a billion makes his story all the more amazing.

Muhammad created a society built on religion, helping others, and equality. Arabia had never seen anything like it. People became united under Muhammad by a belief in one God and not based on ethnic and tribal affiliations. Allied with these early Muslims were Jews, Christians, and pagans as well, creating a multifaith realm. Arabia had been in perpetual war—until

Muhammad unified Arabia in peace. The death and destruction common in Arabia yielded to a diverse and peaceful community. America fulfills this diversity, too, where people of different backgrounds and social classes come together, unified in a democratic system. These ideas are still developing in our world today. A majority of the world's people are still striving for these freedoms, including other Muslims.

"How will you be opening your fast today?" the radio host asked the other guest. I was on the telephone doing a radio interview from my apartment in New York. It was the winter of 2001, and my first book had been published that previous November. It was during *Ramadan,* and the other Muslim guest, who was in a studio in California, and I were talking about *Ramadan* and its meaning. While the other guest was talking about the *iftaar* he would attend at his local mosque, meaning the meal and prayer service held at sunset, I was racking my brain. What should I say? I hadn't fasted that day, but I had a good reason for it.

"So, Asma, how will you be opening your fast?" It was the moment of truth. So I told the truth.

"I won't be, actually, because I'm menstruating. The Qur'an says that women who are menstruating should not fast but make up those days later on after *Ramadan* ends," I said in my usual, exuberant tone.

I think I regretted what I said the moment I said it. But I couldn't lie. Wasn't it worse to lie about fasting than to tell the truth about menstruating?

"Oh!" blurted the host. I had definitely taken him by surprise. "And you're traveling too, right?" he said. "People who are traveling do not have to fast either and can make up the days," he quickly added, wanting to wash the thought of my menstruating off his mind probably.

"Yes, that is true," I said politely. I decided not to say anything else.

The interview soon ended, and the publicist who had arranged the interview called me immediately after. "When I told you to be open," he said, "I didn't mean *that* open." It was one of my first interviews ever, and I felt like I had made a huge faux pas. "It wasn't so bad," the publicist said. "But you probably caused a few fender benders." *Yikes,* I thought. I immediately called my mom and dad for consolation (or punishment, depending on how they took it).

"Asma," my father said. I was expecting something harsh. "You should not feel bad," he chimed. "You should be glad! The Qur'an talks about menstruation. I think even the Prophet Muhammad talked about it," my father continued. "If Islam is not ashamed of it, you shouldn't be either," he concluded.

"Let people hear that Islam talks about menstruation," my mom said boldly, who was with my dad on speakerphone. "Islam doesn't have a problem with it—people do," she added in an agitated and impatient tone.

In Islam, very few topics are off limits. Islam addresses practically every problem or dilemma—either in the Qur'an or in the records of the life of the Prophet Muhammad. In Islam, Muslims are directed to four sources to figure out how to practice Islam properly. First is the Qur'an, which lays out all the major rights and themes. The next source is the *sunna* and *hadiths* of the Prophet Muhammad. The *hadiths* are the sayings and quotations attributed to Muhammad, which provide direction in a specific situation. The *sunna* includes *hadiths* and other recorded actions of the Prophet—such as what kind of clothes he wore or the hygiene he practiced. I compare the *hadiths* to the Bill of Rights. Though not as important as the Qur'an, the *hadiths* help us interpret the Qur'an. (Not all *hadiths* are created equal. Some are fake or manufactured and not said by the Prophet Muhammad at all. Scholars have published lists of the more authoritative ones, an evaluation they came to by studying the *isnad* or chain of transmission to discern the reliability of those who passed the saying down over time.)

With this foundation of sources to call upon, Muslims are able to address the problems they face without feeling ashamed or thinking that the topic is inappropriate to be discussing in a religious context. American culture has a similarly open attitude because we, as Americans, also have such a foundation—with the Constitution and Bill of Rights espousing freedom of speech, as well as our history of dissent and political expression. One of America's most treasured freedoms—freedom of speech—has a home in Islam too.

However, even the Qur'an and the *hadiths* may not have all the answers. The world is very different now than Arabia was 1,400 years ago when the Qur'an was revealed and the *hadiths* are supposed to have taken place. Where the advice of the

Qur'an and the *hadiths* still leaves a gray area, Muslims can rely on the consensus of scholars—called *ijma* in Arabic—and also on analogy—called *qiyas* in Arabic—to interpret the Qur'an as well as a third principle called *shura,* or consultation. Consensus requires that scholars, representing the world Islamic community, all agree on an interpretation. Once they agree, the interpretation becomes a formal practice in Islam. However, all must agree. The moment one member of the community disagrees, the solid consensus begins to crumble, and that aspect of Islam is no longer required of Muslims. Analogy allows Muslims to analogize their current situation to one described in the Qur'an or *hadiths*. Different schools of thought debate whether previous analogies are binding or each analogy is only useful for the situation it was devised for. The final source, consultation, is basically a decision-making process where scholars consult with one another to reach an understanding of the Qur'an.

These various Islamic sources reflect the way Americans are encouraged to think about problems. A writing instructor of mine in law school once said that the most useless opinion to cite in an argument before a court is the minority opinion from a court of a different jurisdiction on a point that was not crucial to the outcome of the case. American lawyers are taught to use similar precedents, to reason and to analogize. Similarly, when judges all agree on a matter of law or an interpretive approach, that aspect becomes accepted by all. If that interpretation is challenged successfully, it becomes the source of debate. These recognized ways of interpreting American law have been practiced by Muslims for centuries in Islamic history. So, for a Muslim, the process of how a law is interpreted is very similar to how Islam has, over time, evaluated a variety of scenarios. Like the American legal system, Islam does not resist questions and dilemmas but fully addresses them. Both Islam and American law are stronger for this built-in attitude.

I am sometimes criticized quite angrily for these views. About a year ago, I was taking questions after giving a speech at Metro State College in Denver, when a young man stood up and yelled, "How can you say that you are a Muslim and an American both first? America means nothing to me."

"Nothing," he repeated in a foreign accent I couldn't place. As is often the case during these Q & As, the questioner was more interested in delivering a political tirade than in engaging in dialogue. "My Islam comes first!" he added militantly. His cronies and some other Muslims who agreed with him applauded loudly as he stared at me from the microphone set up for questioners. I wanted to say, "Are you done yet?" or "I have no response to that," but I figured he wasn't up for a joke.

I had just explained in my speech that a Muslim is instructed by the Qur'an to be kind, devout, and family oriented. The Qur'an encourages Muslims to have children and to educate them. Islam is focused on faith, family life, improving the society one lives in, the protection of property rights, and hard work. These values are American, too, I had said. After the horrifying attacks of 9/11, many people asked me if I was Muslim first or American first. I told the audience in Denver that I am both first at the same time—not just because I don't find that the two conflict with each other but also because they often overlap. Being Muslim makes me a better American, and being American makes me a better Muslim.

"I disagree," I said to the incensed young man. "Islam advises us to follow the laws of our country, and I don't think anything in Islam prevents me from being an American," I responded, knowing that nothing I would say was going to convince him. In hindsight, I realize I should have said, "You should leave America then. Why stay in a place that doesn't matter to you?"

After I do a television appearance or an essay of mine is published, I often receive an angry e-mail through my Web site. In ALL CAPS, the writer will generally ask, "HOW CAN YOU SAY AMERICA IS NOT AT WAR WITH ISLAM? WE ARE, AND YOUR SIDE WILL LOSE!" Both the young man and e-mail writers assume that I am on a different side than either of them. But if I am not with this young Muslim man and clearly not with the jingoistic e-mail writer, who am I with?

To My Critics

Many people will disagree with this entire chapter—and within this group will be both Muslims and Americans. The Muslims will say their allegiance is to Islam only and that America means nothing to them, as the young man did at my speech. They will say that America is out to destroy Islam and that America only cares about money and material things. To them I say, move to an Islamic country then. If you don't need America, then we don't need you. Furthermore, if America is so bad, then why are you even here? What possibly could be keeping you in this land you think is so horrific? Of course, no place is perfect, but no place, even Islamic countries, comes close to America.

The critical Americans will say that Islam and America are opposed to each other and that we are at war. (They actually sound a lot like the zealot Muslims.) To them I say, if America is truly at conflict with Islam, then America is at war with itself too. The values at the heart of Islam—social justice, gender equality, racial equality, property rights—are supposed to be the core of American values too. If America is fighting with something so similar to it, then America no longer stands for what it's supposed to. Today, America comes the closest to fulfilling the ideal state as described in the Qur'an, where one is free from tyranny and democracy reigns. Of course, Muslims do not practice Islam perfectly, and Americans don't necessarily practice American values perfectly. But we have to respect that we are all trying to do the best we can. To focus on the shortcomings is just taking more energy from doing better. That is all that God and our country ask of us. I'd say that we are doing pretty well too.

So maybe everyone was not sitting in his or her proper seats at my sister's wedding. So I was famished, and so we came in the wrong door. I am pleased to report that my sister is indeed happily married. Whether the happily married ladies' good fortune rubbed off on Aliya or not, we'll never know. I do know, however, that bona fide happily married women accompanied her.

As an American Muslim woman, I am free to decide what I want my own wedding to be like. As long as I meet the requirements the Qur'an lays out—which are quite simple, such as the bride must consent, the groom must provide a consideration or dower to the bride, and so on—I can have the happily married ladies if I like. I've decided that if I do have a traditional South Asian wedding, I would like to be accompanied by happily married career women. I'll make my own tradition—one that embodies *my own American Muslim ethnic culture.* You see, Islam allows for my full identity.

Bibliography

I read or reread the following books for inspiration, ideas, translations of the Qur'an and Sufi poetry, and for general information. Where appropriate, I cited them. I wanted to include a list here so that readers can find these wonderful books, too.

Ahmed, Salahuddin. *A Dictionary of Muslim Names.* New York: New York University Press, 1999.

Ali, Abdullah Yusuf. *The Holy Qur'an: Text, Translation, and Commentary.* Elmhurst, NY: Tahrike Tarsile Qur'an, 2001.

Armstrong, Karen. *A History of God: The 4,000-Year Quest of Judaism, Christianity and Islam.* New York: Ballantine, 1993.

———. *Islam: A Short History.* New York: Random House, 2000.

Asad, Muhammad. *The Message of the Qur'an.* Gibraltar: Dar Al-Andalus, 1980.

Bentounès, Sheikh Khaled. *Sufism: The Heart of Islam.* Translated by Khaled El Abdi. Prescott, AZ: Hohm Press, 2002.

Ernst, Carl W. *Following Muhammad: Rethinking Islam in the Contemporary World.* Chapel Hill: University of North Carolina Press, 2003.

Hasan, Asma Gull. *American Muslims: The New Generation.* New York: Continuum, 2000.

Hassan, Rifaat. "Women in the Context of Change and Confrontation Within Muslim Communities." In *Women of Faith in Dialogue,* edited by Virginia Ramey Mollenkott. New York: Crossroad, 1987.

Hotaling, Edward. *Islam Without Illusions*. Syracuse: Syracuse University Press, 2003.

Ladinsky, Daniel, trans. *The Gift: Poems by Hafiz, The Great Sufi Master.* New York: Penguin Compass, 1999.

Mafi, Maryam, and Azima Melita Kolin, trans. *Rumi: Hidden Music.* London: Element, 2001.

Makiya, Kanan. *The Rock: A Tale of Seventh-Century Jerusalem*. New York: Vintage, 2001.

Nasr, Seyyed Hossein. *Islam: Religion, History, and Civilization*. San Francisco: HarperOne, 2002.

Newby, Gordon D. *A Concise Encyclopedia of Islam*. Oxford: Oneworld, 2002.

Rosen, Lawrence. *The Culture of Islam: Changing Aspects of Contemporary Muslim Life*. Chicago: University of Chicago Press, 2002.

Schimmel, Annemarie. *My Soul Is a Woman: The Feminine in Islam.* Translated by Susan H. Ray. New York: Continuum, 1997.

Wadud, Amina. *Qur'an and Woman: Rereading the Sacred Text from a Woman's Perspective*. Oxford: Oxford University Press, 1999.

Wilcox, Lynn. *Women and the Holy Qur'an: A Sufi Perspective*. Vol. 1. Riverside, CA: M.T.O. Shahmaghsoudi, 1998.

Acknowledgments

Being published by HarperOne is a dream come true for me. I wouldn't be here if it weren't for Jana Riess, who is the most generous friend I have ever had and the nicest person I know. Thank you to Mark Tauber of HarperOne for taking me on! Thanks also to Jan Baumer, my editor.

The most enthusiastic thank you and gratitude to my parents, Dr. and Mrs. Malik and Seeme Hasan, who are responsible for a lot of my success! My mom has read almost every draft of everything I have written, no matter how rough, and her and my dad's support are immeasurably important.

My sister, Aliya, was essential to the first version of this book. She did research for me and has always believed in my message. Most of all, she married my dear brother-in-law, Rehan, who has been a true source of friendship and support. His advice and calm guidance helped me renew my efforts on this book. My brother, Ali, was always ready with confident advice and read over the manuscript "last minute."

Dr. Maher Hathout, of the Muslim Public Affairs Council, has been my personal authority on the Qur'an, always generous with his personal time and patient with his explanations. Professor John Esposito went to great effort to review the manuscript for me—an honor in itself.

I want to thank some unofficial members of Team Asma, who played a major role in the first version of this book: my former agent, Tom Grady; my original editor, Greg Brandenburgh; and my "book midwife," Caroline Pincus.

Sincere thank you to my first editor, Frank Ovies, who "discovered" me, and my lawyer and friend, Glenn Merrick.

Special thanks to my friends, family, and personal support staff. You know who you are!

This book was written on a Mac.

—A.G.H.

About the Author

Asma Gull Hasan is the author of *Why I Am a Muslim* (HarperCollins Thorsons/Element, 2004) and *American Muslims: The New Generation* (Continuum, 2000). The daughter of Pakistani immigrants and born in Chicago, Hasan's writing has been described as "a groundbreaking portrait" of the growing American Muslim community (*Christian Century* magazine).

Her second book, *Why I Am a Muslim,* was nominated in 2005 by the National MS Society for its 2005 Books for a Better Life Award. The State Department has invited Hasan to lecture on behalf of the U.S. State Department all over the world, talking about Islam in America. Her books have been translated into French and Japanese and are distributed all over the world.

Her op-eds have been published in the *New York Times,* the *San Francisco Chronicle,* and Beliefnet.com among many others. She was a columnist for the *Pueblo Chieftain,* the *Denver Post,* and the *Pakistan Link* newspapers. She is also a frequent guest on the FOX News Channel, including *The O'Reilly Factor,* and has also appeared on *Today, Anderson Cooper 360, Politically Incorrect* with Bill Maher, *Fresh Air* with Terry Gross, and the History Channel.

She recently finished a special year-long blog for *Glamour* magazine on the 2008 presidential election.

Hasan is a graduate of the New York University School of Law; Wellesley College, *magna cum laude* and a Durant Scholar; and the Groton School, *magna cum laude*.

The author's Web site is www.asmahasan.com.